W9-AOA-070

A Practical Guide to Enzymology

Biochemistry

A Series of Monographs

Editor: ALTON MEISTER, Cornell University Medical College, New York

Volume 1	Woon Ki Paik and Sangduk Kim	PROTEIN METHYLATION
Volume 2	Philipp Christen and David E. Metzler	TRANSAMINASES
Volume 3	Clarence H. Suelter	A PRACTICAL GUIDE TO ENZYMOLOGY

A Practical Guide to Enzymology

Clarence H. Suelter

Department of Biochemistry
Michigan State University
East Lansing, Michigan

A WILEY-INTERSCIENCE PUBLICATION

JOHN WILEY & SONS
New York • Chichester • Brisbane • Toronto • Singapore

Library of Congress Cataloging in Publication Data

Suelter, Clarence H., 1928–
A practical guide to enzymology.

(Biochemistry (John Wiley & Sons),
ISSN 0194-0538 ; v. 3)
"A Wiley-Interscience publication."
Includes bibliographies and index.
1. Enzymes—Analysis. I. Title. II. Series.
[DNLM: 1. Enzymes. W1 BI635B v.3 / QU 135 S944p]
QP601.S84 1985 574.19′25 85-9465
ISBN 0-471-86431-5

Printed in the United States of America

10 9 8 7 6 5 4 3 2 1

To
my parents
and
my teachers

Preface

Enzymology continues to develop as an important and exciting subdiscipline of biochemistry. Enzymologists today, in addition to studying the function and control of enzymes in a metabolic scheme, are placing more and more emphasis on their industrial and medical applications. Yet, regardless of their goals, it is still necessary to first purify enzymes to homogeneity so their characteristics can be assessed without interference from contaminating proteins. Consequently many procedures and techniques have accumulated over the past 60 years, since Sumner first crystallized urease, for denaturing and stabilizing proteins (Chapter 1), quantitating and assaying enzymes (Chapter 2), purifying enzymes (Chapter 3), and characterizing them (Chapters 4 and 5).

The purpose of this book is to bring the practical elements of experimental enzymology together in one small volume. It is intended for those who have completed a one-semester course in biochemistry and who wish or need to develop an expertise in working with enzymes. In addition, it summarizes considerable information that should be of use to the more established investigator. While the book does not provide a comprehensive review of protein structure and function, one of its purposes is to provide a sense of the diversity of the experimental approaches to studying enzymes. Yet much of what is known about proteins, including enzyme mechanisms and their study, is not considered. Also because of the availability of several excellent enzyme kinetics texts, only the introductory aspects of this important area of enzymology are discussed. It is also unfortunate that space does not allow a discussion of the fascinating history of enzymology.

It is impossible to prepare a book such as this without the help of many friends and colleagues. First, I wish to acknowledge the assistance of graduate students and postdoctoral fellows who over the past 20 years have made my laboratory a stimulating environment in which to work. In addition, I owe much to H. David Husic, Jeffrey Baxter, Vickie Bennett-Hershey, Stephen Brooks, and Peter Toth for reading most of the text and for their insight into problem areas. I also wish to recognize others who have graciously provided comments, specifically, Shuyn-long Yun, William D. McElroy, John Wilson, Shelagh Ferguson-Miller, Carol Epstein, Marlene DeLuca, and Robert Barker. Finally, I wish to thank my daughter, Jenifer Nelson, for providing excellent grammatical criticism, my son, Kevin, for preparing most of the illustrations, and my wife, Loretta, for typing the manuscript. Their patience over the last three years enabled me to complete the manuscript. While I am indebted to all of the above for most of the successes that this book may achieve, I must accept full responsibility for the errors of both omission and commission. I would be obliged if I could be informed about errors and omitted techniques and methodologies that are useful in the study of enzymes.

CLARENCE H. SUELTER

East Lansing, Michigan
February 1985

Contents

A Practical Guide to Enzymology

1

Protein Structure and Stability

Proteins are composed of amino acids linked together by peptide bonds. Each polypeptide chain with its unique amino acid sequence folds into a compact three-dimensional protein molecule. The earliest protein structures solved by X-ray crystallography showed the nonpolar amino acids buried in the interior of the molecule and the majority of the polar amino acids on the surface exposed to the solvent. Recently, though, a more detailed quantitative examination of the amino acids on the surface of the protein showed this to be an oversimplification (1). Up to one half of the exposed surface groups are apolar. This evidence is consistent with other data showing that no one single force predominates in determining the final three-dimensional structure (2): The biologically active native structure of a protein is determined by a highly specific pattern of hydrogen bonding, electrostatic interactions, and hydrophobic forces.

Studies by several investigators [see Reference (3) for a review] show that 75% of the interior space of most proteins is filled. In contrast, water and cyclohexane have 58 and 44%, respectively, of their space filled with atoms. Yet, despite the compactness of a protein molecule, it is flexible and undergoes dynamic internal fluctuations (4, 5). For example, the exchange rates of the interior peptide hydrogens of native proteins with the protons of the solvent H_2O are orders of magnitude slower than the exchange rates of the peptide hydrogens in the unfolded protein molecule; yet most of the hydrogens in the interior of the protein are still accessible to the solvent. Apparently the protein fluctuates rapidly between numerous loosely related conformational states forming a dynamic conformational equilibrium.

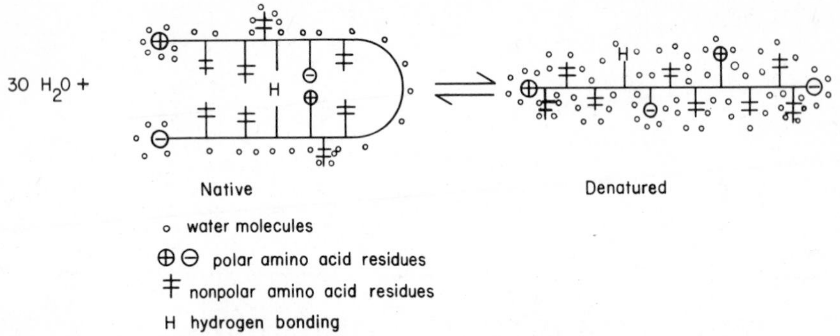

$30\ H_2O$ +

Native

Denatured

o water molecules

⊕⊖ polar amino acid residues

‡ nonpolar amino acid residues

H hydrogen bonding

Fig. 1.1. Schematic diagram of folded native protein molecule in equilibrium with its unfolded denatured form.

This chapter focuses on the following question: *How do different solvents or environmental conditions affect the dynamic three-dimensional structure of proteins?* An appreciation of how different salts and solvents affect protein structure is basic to developing effective laboratory skills for the practical use of enzymes. Some salts and solvents stabilize proteins, some denature* proteins, and some are more effective than others in eluting proteins from chromatographic columns. Finally, an appreciation of protein structure is essential for understanding the physical, chemical, and biological properties of proteins.

THE ROLE OF WATER

"The molecular events which make up life processes have as their common substrate water—the only naturally occurring inorganic liquid." This statement by Franks and Eagland (6) sets the stage for our discussion of the effect of a solvent on protein structure and function. Our task is to examine how various salts, organic solvents, and physical forces affect the ability of water to solvate and thereby influence the three-dimensional structure of a protein in solution.

The effect of different solvating conditions on the three-dimensional structure of a protein molecule is easier to understand when examined in terms of their effect on the equilibrium between the folded native and unfolded denatured forms of a protein in aqueous solution, as depicted in Figure 1.1. The small circles in Figure 1.1 denote water molecules

* See p. 12 for a definition of denaturation.

associated with the protein. Note that the interior of the folded native protein molecule is not solvated by water and that electrostatic interactions, hydrogen bonds, and nonpolar forces participate in forming the folded, native form. The majority of the amino acids in the unfolded denatured protein are completely solvated by water. Thus, compounds or conditions that promote solvation of the amino acid residues by water shift the equilibrium of the reaction in Figure 1.1 toward the unfolded denatured forms; compounds or conditions that reduce the tendency of water to solvate the amino acid residues shift the equilibrium toward the folded biologically active form.

Earlier, it was said that no one single physicochemical force, that is, hydrogen bonding, electrostatic interactions, or hydrophobic forces, determines the final three-dimensional structure of most protein molecules. Yet much more emphasis is placed on nonpolar hydrophobic interactions. Rarely will the effect of solvent on hydrogen bonding and electrostatic interactions be discussed, because the majority of these forces only become important as the polypeptide chain folds upon itself during formation of the biologically active protein. As the nonpolar interactions develop, water is excluded from the interior of the protein molecule and is prevented from competing for the hydrogen bonding and electrostatic interactions. Of course, the attractive or repulsive forces of charged surface residues are also important in defining the final three-dimensional structure, particularly at low ionic strength.

The exclusion of water from the protein interior, as it folds upon itself, is an important aspect of the force involved in maintaining the final three-dimensional structure. It appears that the interaction of parts of the protein molecule with itself is stronger than the interaction of these parts with water. Actually, it is known from thermodynamic measurements that the interaction between nonpolar amino acid residues is due to the strong cohesive effects of the hydrogen bonded water molecules, which forces nonpolar molecules (or parts of nonpolar molecules) to interact with each other regardless of whether or not there is a net attractive force between them (7). These observations together constitute the classical model of hydrophobic bonding.

THE EFFECTS OF SPECIFIC LIGANDS BINDING TO PROTEINS

Compounds or ligands, such as substrates, metal cofactors, and allosteric effectors, that interact with specific sites on proteins (enzymes), exert their effect on protein structure at concentrations that are normally not large enough to affect the structure of the solvent. The formation of these protein-specific ligand complexes either enhances or reduces protein stability. The common inorganic phosphate buffer may also interact with specific sites on some enzymes, particularly those with phosphorylated substrates (8). Posttranslational covalent modifications, such as glycosylation, phosphorylation, and acetylation (see Chapter 4), also affect protein structure; again, they may either enhance or reduce protein stabilty. It is not possible to predict a priori how specific ligands or posttranslational modifications will affect protein stability.

GENERAL SALT EFFECTS

Salts, depending on their concentration, primarily affect electrostatic and nonpolar interactions. At concentrations less than 0.1 *M*, salts function mainly by modifying the nature of surface charges. For example, proteins not soluble in low ionic strength buffers are subjected to surface electrostatic forces, which promote three-dimensional structures that are not solvated sufficiently to prevent aggregation. These electrostatic forces are neutralized by the addition of 0.1–0.2 *M* salt.

Salts at concentrations exceeding 0.2 *M* not only neutralize the electrostatic forces on the protein surface, but also interact with protein structural elements or modify the structure of the solvent, or both, and thus affect the final three-dimensional structure and stability of proteins. Several different approaches have been used to study the effect of salts on the structure of proteins. Von Hippel and Wong (9) examined the effect of increasing salt concentrations on the change in the midpoint temperature T_M of a thermally induced transition of ribonuclease. Their data, provided in Figure 1.2, show that KH_2PO_4 and $(NH_4)_2SO_4$ increase T_M, that is, these salts tend to protect the protein against thermal denaturation while KCl and NaCl have very little effect on T_M. Urea and guanidinium (Guan) salts decrease the T_M of ribonuclease (Figure 1.3). Their effectiveness as denaturing agents decrease in the following order:

$$\text{GuanHSCN} > \text{GuanHCl} > \text{GuanHAc} > \text{urea} > (\text{GuanH})_2\text{SO}_4$$

The influence of high concentrations of electrolytes on protein structure can also be investigated by examining their influence on the solubility

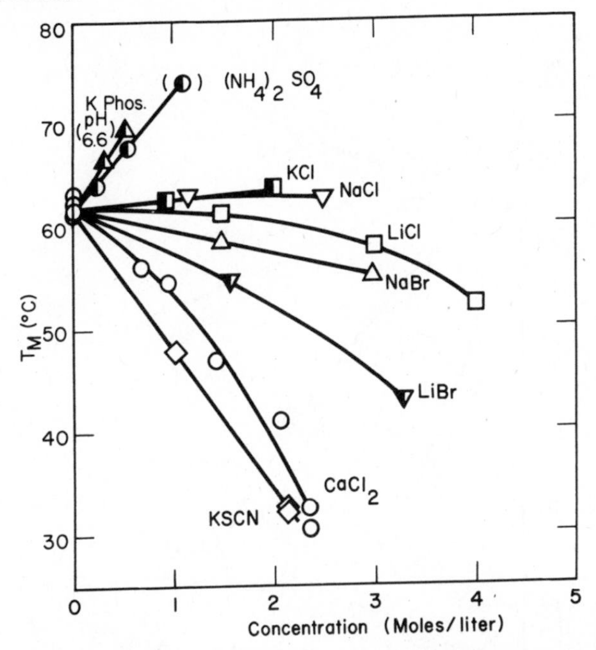

Fig. 1.2. Transition temperatures of ribonuclease as a function of concentration of various added salts. All of the solutions were adjusted to pH 7.0 and contained 0.15 *M* KCl, 0.013 *M* sodium cacodylate, and ribonuclease, 5 mg/mL. Adapted by permission of the copyright owner, The American Society of Biological Chemists, Inc. Reference (9).

of nonpolar amino acids. Returning to the reaction depicted in Figure 1.1, one would predict that compounds that increase the solubility of nonpolar amino acid residues should solvate the interior of a protein molecule and so decrease the stability of the protein. Conversely, compounds that decrease the solubility of nonpolar amino acid residues should stabilize the protein. Data from several studies (9–11) confirm our predictions from the equilibria depicted in Figure 1.1. Salts that enhance the solubility of nonpolar amino acids promote denaturation of protein molecules. Salts that decrease the solubility of nonpolar amino acids promote formation of the three-dimensional structure of proteins, that is, they stabilize proteins.

Interestingly, the solubility of sickle-cell deoxyhemoglobin exhibits the same behavior as nonpolar molecules when placed in solutions of increasing concentrations of various salts (12). The effect of salt on the equilibrium of the reaction

$$n\text{Hb} \rightarrow (\text{Hb})_n$$

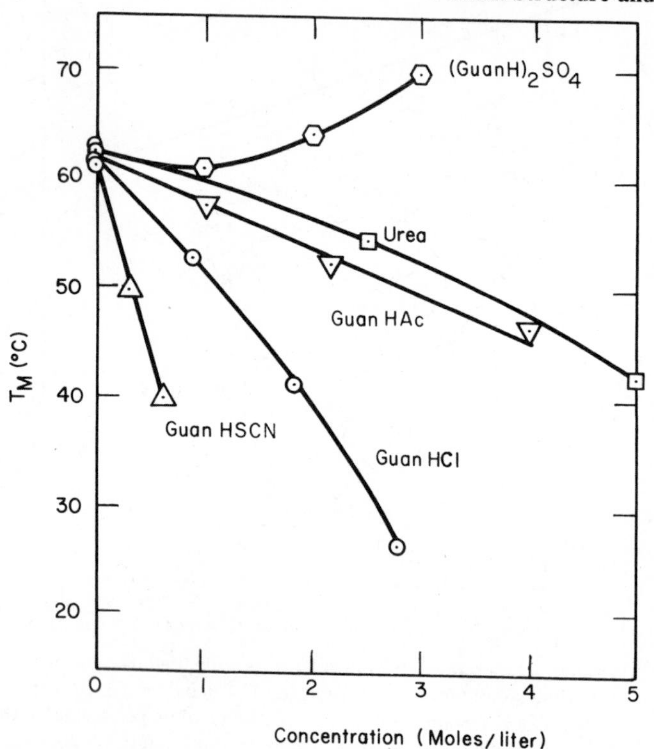

Fig. 1.3. Transition temperatures of ribonuclease as a function of concentration of urea and various guanidinium salts. All solutions were adjusted to pH 7.0 and contained 0.15 *M* KCl, 0.013 *M* sodium cacodylate, and ribonuclease, 5 mg/mL. Adapted by permission of the copyright owner, The American Society of Biological Chemists, Inc., Reference (9).

where nHb is soluble hemoglobin and $(\text{Hb})_n$ is insoluble hemoglobin, can be described by the Setschenow equation

$$S_0/S = K_s/C_s$$

where K_s is the Setschenow constant or the salting out constant, and C_s is the concentration of a salt. The ratio of the solubility of a substance in the absence of salt to that in the presence of salt is S_0/S. Compounds that increase the solubility of deoxyhemoglobin *S* and nonpolar amino acids, that is, decrease K_s, also lower the transition temperature for unfolding ribonuclease. Compounds that decrease the solubility (salt out), raise the transition temperature.

AQUEOUS ORGANIC SOLVENTS

Aqueous organic solvents were used extensively by early enzymologists to fractionate proteins. Payen and Persoz (13) first used ethanol to precipitate diastase. Sumner (14) used acetone to purify urease and Cohn et al. (15) developed procedures using ethanol to prepare globular protein fractions from blood serum. Many other enzymes have been purified since these early studies, by fractional precipitation with aqueous ethanol, methanol, or acetone.

The use of aqueous organic solvents to fractionate proteins declined after the development of chromatographic techniques. Around 1965, however, there was a renewed interest in the use of organic solvents, both to purify proteins and to solve a variety of other enzymological problems. For example, polyhydroxylic compounds such as glycerol stabilize proteins, particularly against cold denaturation; poly(ethylene glycol), (PEG) which is a polyether rather than an alcohol, is effective in precipitating proteins, and as discussed in Chapter 3, detergents and organic solvents are needed to purify integral membrane proteins. Finally, a variety of organic alcohol–water mixtures are used in cryoenzymology, the purification and study of proteins at temperatures ranging from −20 to −60°C.

Monohydric Alcohols

It is common knowledge that high concentrations (30%) of ethanol denature proteins. Nonpolar alcohols such as ethanol replace nonpolar interactions between the interior amino acid residues by interacting directly with them (16). In 1967, however, Brandts and Hunt (17) discovered that at 0–10°C, ethanol at concentrations less than 20% acts as a strong stabilizing agent for ribonuclease. Low concentrations of monohydric alcohols have since been used to stabilize proteins. For example, 1% by volume concentrations of either methanol or ethanol accelerate the rate of renaturation of dissociated tetrameric L-asparaginase (18); glyoxylase I is stable for at least 11 months at 4°C when stored in 10% aqueous methanol (19) (it is not stable in the absence of methanol). Low concentrations of organic solvents (Table 1.1) are also effective in stabilizing proteins against mechanical shaking (20). However, in this example, the longer chain alcohols are more effective on a molar basis than the short chain alcohols. Concentrations of the solvent greater than 10 × C_{50} (see Table 1.1 for a definition of C_{50}) lead to rapid denaturation at 20°C.

Table 1.1. Effect of various organic solvents on the denaturation of sickle-cell oxyhemoglobin *S* by mechanical shaking[a]

Compounds	$C_{50}{}^{b}$ (mM)	% (v/v)
Methanol	303.4	1.2
Ethanol	130.0	0.75
Propanol	65.8	0.60
Butanol	54.7	0.50
Ether	22.0	0.15
Dioxane	42.1	0.36
Acetone	70.1	0.52
Methyl ethyl ketone	42.0	0.38
Dimethylformamide	45.2	0.35
Hexamethylphosphoramide	0.83	0.015
2,2-Dimethylpropane	12.3	0.15
n-Amyl alcohol	75.9	0.82

[a] Adapted by permission from Reference (20).
[b] Indicates concentration of alcohol required to reduce the first-order rate constant for denaturation by a factor of two.

One of the major benefits resulting from the study of proteins at low temperatures is the development of useful solvent systems for working with proteins under these conditions (21). It is now known that many enzymes possess the same catalytic potential in mixed organic solvents at low temperatures as in aqueous solutions at room temperature. This behavior enables cryoenzymologists to determine the kinetic and thermodynamic constants for the elementary steps of several enzyme catalyzed reactions as well as to stabilize the enzyme–substrate intermediates long enough to permit acquisition of structural information. Cryosolvents can also be used to purify normally labile proteins (22, 23).

Cryoenzymologists commonly use three major types of organic cosolvents: alcohols (methanol and ethanol), polyols (ethylene glycol, glycerol, and 2-methyl-2,4-pentanediol), and aprotic solvents (dimethyl sulfoxide and dimethylformamide). In choosing a suitable cryosolvent, it is important to realize that (a) each solvent mixture affects the pK_a of buffers differently, (b) the viscosity should be low in order to avoid diffusion control of reaction rates, and (c) the dielectric constant should be the same as water. Tables showing changes in the pK_a of various buffers and changes in the dielectric constant and viscosity of aqueous organic cosolvents as a function of temperature are available (24, 25).

A major problem in cryoenzymology is protein solubility. Because each protein is different, some trial and error is necessary in order to find a suitable solvent system.

Polyhydric Alcohols

In contrast to monohydric alcohols, which at concentrations exceeding 10% destabilize proteins at room temperature (17), several polyhydric alcohols show a stabilizing influence. Glycerol and other polyhydroxylic compounds are used by enzymologists to stabilize the activity of many enzymes, particularly those exhibiting cold lability (26–29). Polyhydric alcohols also protect against thermal denaturation. If the thermal transitions of chymotrypsinogen at pH 2 or 3 are examined in the presence of a variety of polyhydric alcohols (Figure 1.4), several points are evident (30–33). First, the increase in ΔT_M is roughly linear with increasing concentration of the alcohol or sugar. Second, the order of decreasing effectiveness on a weight/volume (w/v) basis is

inositol > mannitol = sorbitol > aldonitol
= erythritol > sucrose (pH 3) > glycerol

Ethylene glycol destabilizes at all concentrations (data not shown). Another study (34) of ΔT_M for the stabilization of ovalbumin in 50% (w/v) sugars at pH 7.0 produces the following order of decreasing effectiveness

glucose > sorbitol > xylitol > mannose > fructose > ribitol > sucrose
> maltose > erythritol > dextran 10 > glycerol

Thus, polyhydric alcohols stabilize proteins against both heat denaturation and cold inactivation. In fact (as emphasized in Chapter 3), 10–25% glycerol is a useful solvent for working with proteins in the laboratory. There is not sufficient space in this monograph to discuss, in detail, the thermodynamics of this stabilization. In general, these solvents bind water or reduce its solvating ability and therefore shift the equilibrium of the reaction (depicted in Figure 1.1) to the native protein structure. Consult References (30–33) for additional details.

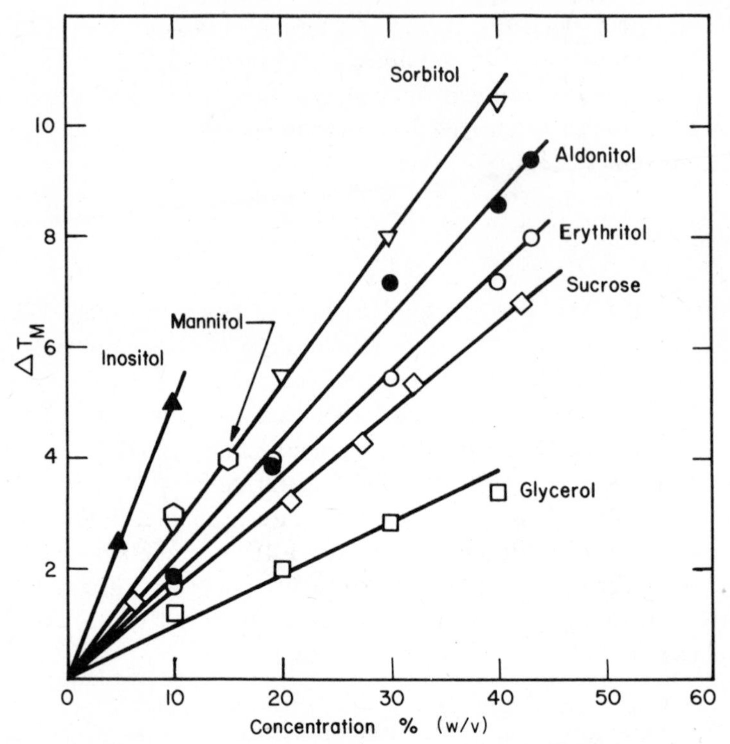

Fig. 1.4. The change in the thermal transition temperature ΔT_M of chymotrypsinogen as a function of increasing concentration of various organic alcohols. The data for sorbitol, mannitol, inositol, aldonitol, and erythritol were obtained with 0.5 mg/mL chymotrypsinogen adjusted to pH 2 with 1 *N* HCl (30). The sucrose (31) and glycerol (32) data were at pH 3, 1 mg/mL chymotrypsinogen.

EFFECT OF PRESSURE AND TEMPERATURE

Pressure

The effects of pressure on enzymes will be discussed in terms of two distinct pressure ranges—nonbiospheric pressures above 3 kbar and biospheric pressures from 1 to 3 kbar. As was first demonstrated by Bridgman in 1914 (35), a pressure of 10 kbar irreversibly denatures proteins. Therefore, high pressure can be used as an alternative to variations in temperature, extremes of pH, or high concentrations of urea or guanidine HCl in protein denaturation studies.

The instability of proteins under pressure is caused by the enhanced solvation of both nonpolar amino acids and charged groups in the hydrophobic interior of the protein molecule. For example, the transfer of methane in hexane to water is accompanied by a $\Delta V = -22.7$ cm^3/mol, and the ionization of a carboxyl group is accompanied by a $\Delta V = -10$ cm^3/mol^1 (36). Relatively small volume changes accompany hydrogen bond formation. Increased pressure then drives the equilibrium of the reaction, depicted in Figure 1.1, to the right.

Oligomeric proteins are more susceptible to pressure induced denaturation than monomeric ones. Consider the following equilibria describing the dissociation of oligomeric proteins promoted by pressure (37).

$$\underset{\text{(active)}}{\text{oligomer}} \xrightleftharpoons{\text{medium pressure}} \underset{\text{(inactive)}}{\text{monomer}} \xrightarrow{\text{high pressure}} \underset{\text{(inactive)}}{\text{unfolded polymer}}$$

The dissociation of lactate dehydrogenase into its subunits by pressure, in contrast to other methods of subunit dissociation, is fully reversible as long as the pressure does not exceed 2 kbar (38, 39). Tetrameric porcine lactate dehydrogenase is half-dissociated at a pressure of 0.76 kbar. On the other hand, monomeric enzymes are characterized by a $P_{\frac{1}{2}} = 4$–8 kbar (40).

A more detailed summary of the effect of pressure on enzyme systems, including conformational changes and ligand binding, is given by Jaenicke (37). One point to realize, however, is that preparative (41) or analytical ultracentrifugation (42) can generate sufficient hydrostatic pressures to affect the structure and stability of some proteins.

Temperature

Most proteins from mesophiles are inactivated by temperatures above 40°C. Proteins from thermophiles, on the other hand, usually withstand elevated temperatures, from 40 to 100°C. For example, glyceraldehyde-3-phosphate dehydrogenase from *Thermus aquaticus*, an extreme thermophile, has a half-life of 30 min at 98°C (43). Extra buried ionic bonds (salt bridges) between subunits and additional hydrophobic interactions amongst loops of the polypeptide chain at the core of multimeric thermophilic proteins appear to account for the enhanced thermostability.

Additional hydrophobic interactions at the edge of the subunit interfaces may also prevent access of water to the interior of the protein molecule (43–45).

A group of oligomeric proteins known as cold sensitive proteins are less stable at 0 than at 25°C (46–48). The inactivation of these proteins at low temperatures, which may or may not be reversible, is often due to dissociation into their inactive monomeric subunits. For example, yeast pyruvate kinase at 0.5 mg/mL, dissociates rapidly into inactive monomers at 0°C; at higher protein concentration the dissociation is much slower ($t_{\frac{1}{2}}$ for inactivation is 75 min at 10 mg/mL) (46). At low temperatures (0 to 5°C) the intersubunit contact forces maintaining the multimeric state of these cold sensitive proteins at room temperature become weaker. In addition to the temperature dependence of the hydrophobic interaction (lower temperatures reduce the entropy of water thus promoting the solvation of the nonpolar amino acid residues and shifting the equilibrium of the reaction depicted in Figure 1.1 to the right), Bock and Frieden (47) suggest that temperature sensitive ionizable groups may also be involved in maintaining the native multimeric state.

Whatever the mechanism of the cold sensitive inactivation, cryoprotectants will usually prevent it. Many insects avoid freeze injury by accumulating such polyols and/or sugars such as glycerol, sorbitol, or trehalose. The larvae of *Bracon cephi* (hymenoptera) accumulate up to 25% of their body weight as glycerol (21). Certain Antartic fish accumulate antifreeze glycoproteins in their blood (48). Common cryoprotectants used in vitro are 20–25% glycerol (21), 10% dimethyl sulfoxide (49), 30–50% ethylene glycol (26), 10% methanol (49), 10% propylene glycol (49), or 1.5 *M* sucrose (50).

DENATURATION AND STABILIZATION

Denaturation of a protein can be defined in a variety of ways. Tanford (51) defines denaturation as any major change of the original native structure of a protein that does not alter the amino acid sequence. Many investigators use loss of biological function as evidence for denaturation. For this discussion, the latter definition is generally assumed. Of more interest, however, are conditions that either stabilize a protein or promote its denaturation.

Denaturation

Often during the course of enzymological studies, enzymes or proteins must be denatured to complete an experiment. Deciding which denaturant to use depends on the purpose of the experiment. When removal of protein, that is, deproteinization is required, one must be concerned with the possible interference of the denaturant remaining in the supernatant in subsequent procedures. If metabolite concentrations in the supernatant are to be determined, the stability of the metabolite and/or adsorption of the metabolite to the protein precipitate must be considered. On the other hand, when rapid inactivation of an enzyme is required then there are other concerns, the most compelling one being the rate at which inactivation is accomplished.

Basically, there are three groups of denaturants (see Table 1.2) used in deproteinizing extracts or reaction mixtures. Perchloric acid is probably the denaturant most used for precipitating proteins. The majority of the perchlorate ion in a clarified extract can be removed as insoluble $KC10_4$ in the cold. The solubility of $KC10_4$ is 54 m*M* at 0°C and 1.6 *M* at 100°C. Stoichiometric amounts of KOH, K_2CO_3, $KHCO_3$, K_2HPO_4, or K_3PO_4 are useful titrants (59). Proteins with M_r = 18,000 or less that may be soluble in 0.5 *N* $HC10_4$ (60) can be precipitated with $HC10_4$ after addition of eight volumes of acetone. If metabolites remaining in the supernatant are to be analyzed, it is important to remove all protein precipitates obtained after treatment with perchloric or trichloroacetic acid before reneutralizing them, since some acid denatured enzymes will recover activity after neutralization.

Proteins in solvents containing Triton X-100 or deoxycholate can be precipitated with 5% trichloroacetic acid after addition of sodium dodecyl sulfate equimolar to Triton X-100 or deoxycholate (61). Trichloroacetic acid can be removed from the clarified extracts by extraction with water-saturated ether; be aware that small molecular weight metabolites may also be extracted (53).

Enzymes or proteins are often denatured (dissociated) as a first step in determining their subunit structure or to inactivate enzymes prior to isolation of metabolites or other macromolecules, such as nucleic acids or polysaccharides. The effectiveness of several denaturants is given in Table 1.3.

Urea is a popular denaturant, even though it is less effective than other denaturants (Table 1.3). A nonionic denaturant like urea in water is useful

Table 1.2. Denaturants commonly used to deproteinize extracts or reaction mixtures

Denaturant	Final Conditions
Acids	
Perchloric	0.3 –0.6 *N*
Trichloroacetic	0.2–0.3 *N*
Tungstic (52)[a]	0.07 *N* H_2SO_4/0.03 *M* Na_2WO_4
Organic solvents	
Ethanol	75–80%, heat to boiling for 1–2 min
Acetone (53)	35–40%
Chloroform/ethanol[b]	0.2–0.4 volume of 1:2.5
Phenol, cresol[c]	
Heavy metals	
$Ba(OH)_2$/$ZnSo_4$[d]	0.05 *M* $Ba(OH)_2$/0.05 *M* $ZnSO_4$
$ZnSO_4$/NaOH[d]	0.1 *M* $ZnSO_4$/0.1 *M* NaOH
Other heavy metals[e]	
Other	
Thermal[f]	
Ultrafiltration[g]	

[a] Tungstic and phosphotungstic acids are no longer used extensively to deproteinize solutions.
[b] Chloroform/ethanol is principally used to remove hemoglobin from lysed erthrocytes (54) or liver extracts (55).
[c] Phenol, cresol, or a mixture is often used for deproteinization during the isolation of ribonucleic acid (RNA) (56). Guanidinium thiocyanate plus 0.1 *M* 2-mercaptoethanol is a more effective denaturant recommended for isolating intact ribonucleic acid (RNA) (57).
[d] Identical amounts of each compound are added for effective deproteinization (58).
[e] Protein precipitates obtained by addition of salts of Cd, Hg, Zn, Cu, or Pb normally adsorb a variety of metabolites (59).
[f] Precipitating proteins by heating is not recommended since precipitation is often not complete and labile metabolites may be hydrolyzed.
[g] Ultrafiltration is an effective but slow procedure for deproteinizing protein solutions.

because proteins or peptides in this solvent can be subjected to ion exchange chromatography without prior dialysis. Denaturation appears to involve the interaction of one urea molecule with each pair of peptide bonds and with each aromatic side chain (65). Because urea slowly dissociates to ammonium cyanate in solution, it should be purified before use (66). Recrystallization from absolute ethanol (67) or passage of a concentrated solution over a mixed bed ion exchange resin (68) is usually

Table 1.3. Effectiveness of useful protein denaturants

	Minimum Molar Concentration Required to Denature			
Denaturant	Aldolase[a]	Lysozyme	Ovalbumin[a]	Serum Albumin[a]
Urea	4.5		8.5	
Guanidinium chloride	3.4	5[b]	5.3	3.4
Guanidinium bromide	2.9		4.7	2.9
Guanidinium iodide	2.3		3.1	2.3
Guanidinium thiocyanate	1.8		2.2	2.0
Carbamoylguanidinium chloride	2.0			2.3
Guanylguanidinium chloride	1.9			2.2
Phenylethylbiguanide[c]		1.25[d]		

[a] Reference (62).
[b] Reference (63).
[c] *N*-(2-phenylethyl)imidodicarbonimidic diamide hydrochloride.
[d] Reference (64).

sufficient. If a urea preparation of higher grade purity is desired, then the method of Prakash et al. (65) may be used.

Guanidinum salts also denature proteins by binding directly to the peptide bonds and the aromatic side chains (69). Again, it is important to use a high grade guanidinium salt (70) for denaturation, particularly if spectrophotometric measurements on the denatured protein solution are planned.

Of the denaturants listed in Table 1.3, phenylethylbiguanide and guanidinium thiocyanate are the most effective. Guanidinium thiocyanate in conjunction with 2-mercaptoethanol is used to prepare native RNA free from protein contamination (57). Guanidine·HCl is not sufficiently effective in denaturing ribonuclease to allow isolation of intact RNA.

Ionic detergents are particularly effective protein denaturants. The most commonly used anionic detergent, sodium dodecyl sulfate, forms complexes with proteins (71) and dissociates them to their constituent polypeptide chains. Cationic detergents also form complexes similar to the anionic detergents but with diminished affinity (72). Nonionic detergents like Triton X-100, on the other hand, do not bind to proteins as the ionic ones do. Most, but not all, enzymes retain complete catalytic activity in Triton X-100 unless its concentration greatly exceeds its critical micelle concentration of 0.2 m*M*. For a discussion of the differences between these detergents, see Tanford (73).

Not all proteins can be denatured by the simple addition of denaturants. For example, bovine erythrocyte superoxide dismutase retains most of its catalytic activity in 10 *M* urea or 4% sodium dodecyl sulfate (74); 0.1 *M* 2-mercaptoethanol or 1 m*M* ethylenediaminetetraacetic acid (EDTA) is required in addition to the denaturants to cause rapid inactivation. Similarly, the enzyme pronase from *Streptomyces griseus* is stable in 6 *M* guanidine·HCl but denatures when the solvent is supplemented with EDTA (75). Carboxypeptidase is also not completely unfolded in 10 *M* urea unless 2-mercaptoethanol is added to reduce the disulfide cross-links (76). Another proteinase, thermolysin, is completely active in 8 *M* urea (77). In summary then, proteins containing disulfide cross-links or alkaline earth metals are usually resistant to denaturation by common denaturants. Therefore, in the absence of specific information, a protein denaturing solvent should contain 1–10 m*M* EDTA, 10–100 m*M* 2-mercaptoethanol, and a 6 *M* guanidinium salt, preferably guanidinium thiocyanate.

When enzymic cleavages are used as the first step in structural studies of a protein, the substrate protein molecule must be unfolded before proteinases can be effective. Many proteinases such as chymotrypsin, trypsin, thermolysin, and the proteinase from *Staphylococcal aureus* are active in 0.5% sodium dodecyl sulfate or 4–8 *M* urea.

Stabilization

As depicted in Figure 1.1, denaturation results from the unfolding and solvation of the polypeptide chain. Therefore, solvents or environmental conditions that prevent unfolding of the protein or solvation of the protein interior stabilize the protein. As noted earlier, specific ligands and salts, such as ammonium sulfate and potassium phosphate, or polyhydroxylic solvents like glycerol also stabilize proteins. Some proteins, extracellular proteases, for example, are prevented from unfolding by disulfide cross-links or Ca^{2+} binding (76, 77). It is interesting also to note that some proteins that are stable to guanidine·HCl have a high arginine to lysine ratio (78).

Several protein modifications—some based on the above observations—prevent denaturation of proteins. Torchilin et al. (79) show a 3 to 21-fold stabilization of chymotrypsin against thermoinactivation by cross-linking surface carboxyl groups with such bifunctional reagents as ethylene- and tetramethylenediamine. This process mimics the formation of disulfide cross-links. Enzymes immobilized by covalent attachment to a variety of supports or entrapped in polymeric gels are more stable to various types of denaturation (80, 81). Other investigators have increased the stability of proteins by guanidination (78) or by acetamidination (82); both modifications result in the introduction of arginine-like groups. The covalent coupling of polyethylene glycol to a protein (83) also provides a protective effect, not only against denaturation but also against proteolytic degradation. Alternatively, the desired enzyme can be isolated from a thermophilic organism, or from a mesophile containing the requisite deoxyribonucleic acid (DNA) message from a thermophilic organism (45).

The loss or denaturation of proteins by adsorption onto glass, quartz, plastic, cellulose, or dialysis membranes (84), can be prevented in several ways. The loss of brain hexokinase and other enzymes by adsorption onto polypropylene tubes, particularly at the air–water interface, is prevented by coating the tubes with bovine serum albumin (85). Cleaning the inner surface of the tubes with ordinary dishwashing detergent, 1:1 chloroform:methanol, 6 *N* HCl, or treating them with 1% dichlorodimethylsilane in benzene does not prevent loss of the enzyme. Glassware precoated with a 1% (w/v) solution of poly(ethylene glycol) 20,000 and dried at 110°C show diminished adsorption of some proteins (86). The loss of RNA polymerase during a chromatographic purification step can be prevented by including 0.1-mg/mL bovine serum albumin in the eluting solvent (87).

Polypropylene or polycarbonate tubes are actually preferred when working with some proteins. For instance, dissociated porcine lactate dehydrogenase at pH 2.5 is irreversibly denatured by glass, presumably by adsorption (88). This loss is diminished when polypropylene or polycarbonate is used. By using buffers containing 0.1% Tween 20 and 1 *M* NaCl, nonspecific adsorption of proteins to Sepharose-bound antigens, presumbly by hydrophobic and ionic interactions, can be prevented (89).

A recent study of the adsorption of luciferase and bovine serum albumin to glass and plastic surfaces indicates that modifying the solvent is more effective in preventing protein losses than modifying the container surface (90). Modification of the container surface by coating it with bovine serum albumin may be effective, but one cannot be assured that the enzyme in solution will not displace the bovine serum albumin from the container surface. The solvent is modified by adding either 50% glycerol, 0.2 m*M* Triton X-100, or 0.1 mg/mL bovine serum albumin. Salt, at least 80 m*M* $(NH_4)_2SO_4$, is required in addition to Triton X-100 to prevent adsorption of luciferase to glass; conveniently, salt is not required for preventing adsorption to glass when 50% glycerol is used as a solvent. In the absence of salt, glycerol and Triton X-100 are almost equally effective in preventing adsorption of proteins to polyethylene or polypropylene surfaces.

Precipitating proteins by adding solid ammonium sulfate occasionally results in denaturation at an air–water interface. Air dissolved in the buffer is salted out from the vicinity of the salt crystal as fine bubbles to create a large air–water interface area, which promotes protein denaturation. To prevent this, simply add a saturated solution of ammonium sulfate (91) rather than the solid.

When proteins are subjected to vigorous shaking, stirring, or any other similar mechanical action, foaming results and proteins are denatured at the air–water interface. As indicated earlier, a variety of dilute organic solvents prevent the denaturation of sickle-cell oxyhemoglobin by mechanical shaking (20). Whether the reduction in surface tension brought about by dilute methanol or ethanol reduces the denaturation by mechanical shaking is not known. However, shear forces encountered by rapid mixing do not appear to denature enzymes (92, 93).

Multimeric proteins, or proteins with cofactors such as zinc or pyridoxal phosphate, are particularly susceptible to denaturation by dilution because they exist in solution in equilibrium with their dissociated forms, as described by Equations (1.1) and (1.2)

$$\text{multimer} \rightarrow n \text{ monomer} \tag{1.1}$$

$$\text{enzyme–cofactor} \rightarrow \text{enzyme} + \text{cofactor} \tag{1.2}$$

Mere dilution of the protein solution shifts the equilibrium of these reactions to the generally inactive forms. Addition of substrate to form the multimeric enzyme substrate complex usually protects against dilution inactivation by shifting the equilibrium of Equation (1.1) to the left; addition of cofactor often favors the associated form Equation (1.2). Changes in temperature or pH will also affect the equilibria. For example, pig heart lactate dehydrogenase, the H_4 isozyme, is partially dissociated at 0.1 mg/mL and 0°C. Increased temperature favors the dissociated form while the cofactor nicotinamide adenine dinucleotide (reduced) (NADH) favors the associated form (94).

Some enzymes, often called sulfhydryl enzymes, require a reduced sulfhydryl residue(s) for optimum catalytic activity. Because sulfhydryl

Table 1.4. Half-life of thiol compounds in solution[a]

Conditions	Half-Life (hr)			
	2-Mercapto-ethanol	Dithiothreitol	Glutathione	3-Mercapto-propionate
pH 6.5, 20°C	100	40	16	7
pH 7.5, 20°C	10	10	9	5
pH 8.5, 20°C	4.0	1.4	1.3	4.5
pH 8.5, 0°C	21	11	8	13
pH 8.5, 40°C	1.0	0.2	0.2	1.6
pH 8.5, 20°C + 0.1 m*M* Cu^{2+}	0.6	0.6	1.2	0.4
pH 8.5, 20°C + 1.0 m*M* EDTA	100	4	70	100

[a] All thiol compounds were dissolved in 0.1 *M* potassium phosphate buffer. Reprinted with permission from *Biochem. Ed.*, **11**, 70 (1983), R. Stevens, L. Stevens, and N. C. Price, *The Stabilities of Various Thiol Compounds used in Protein Purifications*, Pergamon Press, Elmsford, New York.

enzymes and reagents are susceptible to oxidation by molecular oxygen, they should be stored in capped containers under nitrogen, if possible, and in solutions containing a reduced sulfhydryl reagent and EDTA. Ethylenediaminetetraacetic acid is added to complex trace metals such as Cu, Fe, and Zn. These metal ions either bind to the enzyme sulfhydryl residue or catalyze their oxidation by molecular oxygen (95, 96). Table 1.4 is a list of the half-lives of several thiol compounds in solution under a variety of conditions (96). Several points are evident: (a) 2-mercaptoethanol and dithiothreitol are the most stable, (b) the higher the temperature and pH, the shorter the half-lives, (c) 0.1 m*M* Cu^{2+} enhances the rate of oxidation of all sulfhydryl groups, and (d) EDTA increases the half-life of all thiol reagents. Of the several metals tested, all at 0.1 m*M* (Fe^{3+}, Cr^{3+}, Ni^{2+}, Cu^{2+}, Mn^{2+}, Zn^{2+}, and Ag^{+}), only Cu^{2+} affects the half-life of 2-mercaptoethanol.

Of the sulfhydryl reducing reagents listed in Table 1.5, nonionic compounds are preferred since they usually present fewer problems in subsequent studies. For economic reasons, 2-mercaptoethanol is most often used. Normally 1–10 m*M* is sufficient but occasionally, 50–100 m*M* concentrations are necessary to stabilize an enzyme. Monothioglycerol and 2,3-dimercaptopropanol, which are equally or more effective than 2-mer-

Table 1.5. Properties of useful sulfhydryl reducing reagents

Sulfhydryl Reagent	Molecular Weight	Density of Liquid[a]	Molarity of Liquid[a]	Comments
2-Mercaptoethanol[b]	78	1.117 g/mL	14.3 *M*	Volatile
Thioglycerol	108	1.295 g/mL	12 *M*	
Dithiothreitol or dithioerythreitol	154	(Solid)		Strong reducing agents
2,3-Dimercaptopropanol[b]	124	1.239 g/mL	10 *M*	Moderately soluble in water
Tributylphosphine[b]	203	0.81 g/mL	4 *M*	Slightly soluble in water
L-Cysteine	121	(Solid)		
Thioglycolic acid	92	(Solid)		
Glutathione (reduced)	307	(Solid)		

[a] From Reference (97).
[b] Because of low toxicity and offensive odors, use in hood.

captoethanol, are not used extensively because of their expense. Dithiothreitol and dithioerythritol form six membered dithane rings upon oxidation and as a result are stronger reducing agents than the monothiols (98). Usually, 0.1 to 1 m*M* dithiothreitol or dithioerythritol is sufficient to stabilize sensitive sulfhydryl enzymes. Tributylphosphine (99, 100) also effectively reduces disulfides, but it is rarely used to stabilize sulfhydryl enzymes (101); because of its nonpolar character, it may be useful with nonionic detergents.

Several problems are associated with the use of sulfhydryl reagents. They often interfere with the determination of protein concentration (see Chapter 2), and some are effective chelators and inactivate certain metalloenzymes. Since lower concentrations of dithioerythritol and dithiothreitol are needed to protect most enzymes, they interfere less in determining protein concentrations (see Chapter 2). However, these compounds, as well as cysteine, thioglycolate, and 2,3-dimercaptopropanol, are more effective chelators than the remaining monothiols. Zinc metalloproteins are particularly susceptible to inhibition by mercaptans (102, 103). The zinc metalloenzyme, adenosine monophosphate (AMP) aminohydrolase, loses 78% of its activity in 1 hr in 10 m*M* dithiothreitol and 50% in 1 hr in 40 m*M* 2-mercaptoethanol (103). It is important to realize, however, that the oxidation of sulfhydryl residues may have physiological significance because some enzymes are regulated in vivo by oxidation and reduction of sulfhydryl residues (104). As a result, it is important to examine the effect of a reducing compound on enzyme function before using it.

Renaturation

The fact that denatured or unfolded proteins can be renatured has been known for years. The earliest studies involved proteins with disulfide cross-links (105). Reduced and denatured ribonuclease contains eight half-cystine residues; after dilution or dialysis of the denaturant and oxidation of the reduced sulfhydryl groups by molecular oxygen, only one of the 105 possible sets of four disulfide bridges is formed. In other words, the folding reaction favors the biologically active form. We now know that under the proper conditions most dissociated and unfolded multimeric enzymes refold to yield nearly 100% recovery of their enzymic activity. Because full recovery of enzymic activity requires the correct folding of the polypeptide chain and the correct assembly of the subunits, the whole hierarchy of structural organization in proteins is determined by the amino acid sequence (106).

The main obstacle to complete recovery of enzymic activity is a side reaction forming noncovalent aggregates. The enhanced stability of these aggregates is probably due to the intramolecular interaction of exposed nonpolar surfaces usually buried in the interior of the native molecule. Inactive aggregates can be removed by centrifugation, ultrafiltration, and/or gel permeation chromatography; they can then be dissociated and renatured (107, 108).

To achieve extensive recovery of the activity of a dissociated, unfolded protein requires optimizing the renaturing conditions, that is, temperature, pH, buffer, ionic strength, solvent, protein concentration, and specific ligand concentration. Renaturation can usually be accomplished at any temperature consistent with a stable end product, but good recoveries are more likely to be achieved at temperatures ranging from 0 to 30°C; most reactions are performed at 20–30°C. The temperature that gives optimum yields of the correctly folded enzyme must be defined for each reaction.

The pH of each renaturation reaction must also be defined. There is no systematic study available that indicates optimum recovery at a particular pH such as the pI. A good recovery should be obtained at the pH of optimum stability. Many types of buffers are used: A good recovery, however, requires that the salt concentration be controlled around 0.1–0.2 *M* ionic strength.

Except for the renaturation of metalloproteins and proteins containing disulfide cross-links, all renaturation solvents should contain a sulfhydryl reducing agent such as mercaptoethanol (10–50 m*M*), dithioerythritol (1–10, m*M*), dithiothreitol (1–10 m*M*), glutathione (10–50 m*M*), or thioglycerol (10–50 m*M*) as well as a chelator such as 1–10 m*M* EDTA. Glycerol (25%) (88, 109) or sucrose (0.5 *M*) (110) is required for optimum renaturation of some proteins.

The protein concentration in the renaturation reaction is particularly critical. As a rule, one expects decreased recoveries at both low (<5 μg/mL, and high (>100 μg/mL) concentrations. At low concentrations, the decreased recovery may be caused by surface effects or not allowing sufficient time for reactivation. Recoveries at low protein concentration may be improved by including 1 mg/mL bovine serum albumin in the renaturation reaction (110). High protein concentration favors aggregation, which competes with refolding and the correct association reaction (108).

In some instances, cofactors are needed to renature a functional enzyme. Horse liver alcohol dehydrogenase is not reactivated (no reassociated monomers are detected) in the absence of Zn^{2+}. Optimum yield of renatured enzyme was found at 10 μ*M* Zn^{2+} (106); excessive Zn^{2+} causes formation of inactive aggregates. Erythrocyte carbonic anhydrase B appears to refold in the absence of Zn^{2+}, but at a much slower rate (111). A cofactor such as nicotanimide adenine dinucleotide (NAD^{+}) may enhance the rate of renaturation of pyridine nucleotide dehydrogenases (112), however, it is not normally required (106).

The renaturation reaction conditions discussed up to this point apply to proteins denatured by acid, urea, or guanidinium salts. Yet sodium dodecyl sulfate is also a common denaturant. Proteins denatured with this detergent can be renatured but the approach is different. Sodium dodecyl sulfate should first be removed by anion exchange chromotography in urea (113), by ion retardation chromatography (114), or by extraction as an ion pair with triethylamine or tributylamine into acetone or heptane (115). In the latter case, the protein is recovered as a precipitate; dissolution of the precipitate into urea or guanidine·HCl followed by dilution often yields the renatured protein. Renaturation of a protein in sodium dodecyl sulfate solution can also be achieved by first diluting it in guanidine·HC1 and then with a renaturing solvent (116). Some proteins, particularly monomeric ones, can be renatured after sodium dodecyl sulfate polyacrylamide gel electrophoresis by simply washing the gel extensively with buffer to allow renaturation in situ; sodium dodecyl sulfate washes out of the gel more rapidly than the unfolded protein (117). Not only is it essential to remove the denaturant as soon as possible after electrophoresis, but it may be necessary to allow the renaturation reaction to run for 24 hr (118). Furthermore, it is important to use a pure detergent because some sources of sodium dodecyl sulfate contain hexadecyl and tetradecyl sulfates, which have a higher affinity for proteins and, as a result, renaturation is retarded. If a lower grade sodium dodecyl sulfate is used, the gels may be extracted with 25% (v/v) aqueous isopropanol before renaturing the proteins (119).

Sections of the polyacrylamide gel containing protein can also be crushed and extracted with buffer, removed from the crushed polyacrylamide by centrifugation, precipitated with acetone, and renatured after dissolution in guanidine·HCl (120). Another method involves direct renaturation of the sodium dodecyl sulfate protein complexes by dilution with an excess of Triton X-100. Recoveries of 1–10% have been reported, however, by varying pH, ionic strength, or other parameters, higher yields may perhaps be obtained (121). Additional details regarding renaturation of proteins in polyacrylamide gels are provided in Reference (122).

REFERENCES

1. B. Lee and F. M. Richards, *J. Mol. Biol.*, **55**, 379–400 (1971).
2. J. L. Finney, B. J. Gellatly, I. C. Golton, and J. Goodfellow, *Biophys. J.*, **32**, 17–33 (1980).
3. T. E. Creighton, *Prog. Biophys. Mol. Biol.*, **33**, 231–297 (1978).
4. C. K. Woodward and B. D. Hilton, *Annu. Rev. Biophys. Bioeng.*, **8**, 99–127 (1979).
5. S. W. Englander and N. R. Kallenbach, *Q. Rev. Biophys.*, **16**, 521–655 (1984).
6. F. Franks and D. Eagland, *CRC Crit. Rev. Biochem.*, **3**, 165–219 (1975).
7. W. Jencks, *Catalysis in Chemistry and Enzymology*, McGraw-Hill, New York, 1969, pp. 417–436.
8. M. Rippo, M. Signorini, and T. Bellini, *Biochem. J.*, **197**, 747–749 (1981).
9. P. H. von Hippel and K-Y. Wong, *J. Biol. Chem.*, **240**, 3909–3923 (1965).
10. P. H. von Hippel and T. Schleich, *Acc. Chem. Res.*, **2**, 257–265 (1969).
11. P. K. Nandi and D. R. Robinson, *J. Am. Chem. Soc.*, **94**, 1299–1308 (1972).
12. W. N. Poillon and J. F. Bertles, *J. Biol. Chem.*, **254**, 3462–3467 (1979).
13. A. Payen and J. F. Persoz, *Ann. Chem. Phys.* (*Paris*), **53**, 73–92 (1833); from English translation by H. Friedmann in *Benchmark Papers in Biochemistry*, Vol. 1, Enzymes, Hutchinson Ross, Stroudsburg, PA, 1981, pp. 119–122.

14. J. B. Sumner, *J. Biol. Chem.*, **69**, 435–441 (1926).
15. E. J. Cohn, L. E. Strong, W. L. Hughes, D. J. Mulford, J. N. Ashworth, M. Melin, and H. L. Taylor, *J. Am. Chem. Soc.*, **68**, 459–475 (1946).
16. S. N. Timasheff, *Acc. Chem. Res.*, **3**, 62–68 (1969).
17. J. F. Brandts and L. Hunt, *J. Am. Chem. Soc.*, **89**, 4826–4858 (1967).
18. S. Shifrin and C. L. Parrott, *Arch. Biochem. Biophys.*, **166**, 426–432 (1975).
19. S. Sellin, A-C. Aronsson, and B. Mannervik, *Acta Chem. Scand., Ser. B,* **34**, 541–543 (1980).
20. T. Asakura, K. Adachi, and E. Schwartz, *J. Biol. Chem.*, **253**, 6423–6425 (1978).
21. P. Douzou, *Cryobiochemistry: An Introduction,* Academic, London, 1977.
22. A. L. Fink and S. J. Cartwright, *CRC Crit. Rev. Biochem.*, **11**, 145–207 (1981).
23. P. Douzou, *Adv. Enzymol. Relat. Areas Mol. Biol.*, **45**, 157–272 (1977).
24. P. Douzou, G. Hui Bon Hoa, P. Maurel, and F. Travers in *Handbook of Biochemistry and Molecular Biology,* G. Fasman Ed., CRC Press, Cleveland Vol. 1, 3rd ed. 1976, p. 520.
25. P. Douzou, *Methods Biochem. Anal.*, **22**, 401–512 (1974).
26. J. Jarabak, A. E. Seeds Jr., and P. Talalay, *Biochemistry,* **5**, 1269–1279 (1966).
27. R. T. Kuczenski and C. H. Suelter, *Biochemistry,* **9**, 939–945 (1970).
28. S. L. Bradbury and W. B. Jakoby, *Proc. Nat. Acad. Sci. USA,* **69**, 2373–2376 (1972).
29. S. O'Neil Houch, *J. Biol. Chem.*, **248**, 2992–3003 (1973).
30. K. Gekko and T. Morikawa, *J. Biochem.* (*Tokyo*), **90**, 51–60 (1981).
31. J. C. Lee and S. N. Timasheff, *J. Biol. Chem.*, **256**, 7193–7201, (1981).
32. K. Gekko and S. N. Timasheff, *Biochemistry,* **20**, 4677–4686 (1981).
33. T. Arakawa and S. N. Timasheff, *Biochemistry,* **21**, 6536–6544 (1982).
34. J. F. Back, D. Oakenfull, and M. B. Smith, *Biochemistry,* **18**, 5191–5196 (1979).
35. P. W. Bridgman, *J. Biol. Chem.*, **19**, 511–512 (1914).
36. W. Kauzmann, *Adv. Protein Chem.*, **14**, 1–63 (1959).
37. R. Jaenicke, *Annu. Rev. Biophys. Bioeng.*, **10**, 1–67 (1981).
38. B. C. Schade, R. Rudolph, H-D. Ludemann, and R. Jaenicke, *Biochemistry,* **19**, 1121–1126 (1980).
39. K. Muller, H-D. Ludemann, and R. Jaenicke, *Biochemistry,* **20**, 5411–5416 (1981).
40. S. A. Hawley, *Methods Enzymol.*, **49**, 15–24 (1978).
41. P. Champeil, S. Buschlen, and F. Guillain, *Biochemistry,* **20**, 1520–1524 (1981).
42. R. Josephs and W. F. Harrington, *Proc. Nat. Acad. Sci. USA,* **58**, 1587–1594 (1967).
43. J. E. Walker, A. J. Wonacott, and J. I. Harris, *Eur. J. Biochem.*, **108**, 581–586 (1980).
44. P. Argos, M. G. Rossmann, U. M. Grau, H. Zuber, G. Frank, and J. D. Tratschin, *Biochemistry,* **18**, 5698–5703 (1979).
45. V. V. Mozhaev and K. Martinek, *Enzyme Microb. Technol.*, **6**, 50–59 (1984).
46. R. T. Kuczenski and C. H. Suelter, *Biochemistry,* **10**, 2867–2872 (1971).
47. P. E. Bock and C. Frieden, *Trends Biochem. Sci.*, **3**, 100–103 (1978).
48. R. E. Feeney and Y. Yeh, *Adv. Protein Chem.*, **32**, 191–282 (1978).
49. D. J. Graves, R. W. Sealock, and J. H. Wang, *Biochemistry,* **4**, 290–296 (1965).
50. M. F. Utter, D. B. Keech, and M. C. Scrutton, *Adv. Enzyme Regul.*, **2**, 49–68 (1964).

51. C. Tanford, *Adv. Protein Chem.*, **23,** 121–282 (1968).
52. S. W. Hier and O. Bergeim, *J. Biol. Chem.*, **161,** 717–722 (1945).
53. L. Arola, E. Herrera, and M. Alemany, *Anal. Biochem.*, **82,** 236–239 (1977).
54. D. Herbert and J. Pinsent, *Biochem. J.*, **43,** 203–205 (1948).
55. R. K. Bonnichsen, *Acta Chem. Scand.*, **4,** 715–717 (1950).
56. K. S. Kirby, *Methods Enzymol.*, **12B,** 87–99 (1968).
57. J. M. Chirgwin, A. E. Przybyla, R. J. MacDonald, and W. J. Rutter, *Biochemistry,* **18,** 5294–5299 (1979).
58. M. Somogyi, *J. Biol. Chem.*, **160,** 69–73 (1945).
59. H. U. Bergmeyer, E. Bernt, K. Gawehn, and G. Michal, in *Methods of Enzymatic Analysis,* Vol. 1 (second English ed.) Verlag Chemie GmbH, Weinhein–Bergstra, Berlin, 1974, p. 179.
60. E. Wedege, *Comp. Biochem. Physiol.*, **70B,** 63–68 (1981).
61. K. C. Retz and W. J. Steele, *Anal. Biochem.*, **79,** 457–461 (1977).
62. F. J. Castellino and R. Barker, *Biochemistry,* **7,** 4135–4138 (1968).
63. C. Tanford, R. H. Pain, and N. S. Otchin, *J. Mol. Biol.*, **15,** 489–504 (1966).
64. M. E. Noelken, *Anal. Biochem.*, **104,** 228–230 (1980).
65. V. Prakash, C. Loucheux, S. Scheufele, M. J. Gorbunoff, and S. N. Timasheff, *Arch. Biochem. Biophys.*, **210,** 455–464 (1981).
66. G. R. Stark, W. H. Stein, and S. Moore, *J. Biol. Chem.*, **235,** 3177–3181 (1960).
67. F. H. White, Jr., *Methods Enzymol.*, **25,** 387–392 (1972).
68. E. Haber, *Proc. Nat. Acad. Sci. USA,* **52,** 1099–1106 (1964).
69. J. C. Lee and S. N. Timasheff, *Biochemistry,* **13,** 257–265 (1974).
70. Y. Nozaki, *Methods Enzymol.*, **26,** 43–50 (1972).
71. J. A. Reynolds and C. Tanford, *Proc. Nat. Acad. Sci. USA,* **66,** 1002–1007 (1970).
72. Y. Nozaki, J. A. Reynolds, and C. Tanford, *J. Biol. Chem.*, **249,** 4452–4459 (1974).
73. C. Tanford, *The Hydrophobic Effect: Formation of Micelles and Biological Membranes,* 2nd ed. Wiley, New York, 1979, pp. 159–163.
74. H. J. Forman and I. Fridovich, *J. Biol. Chem.*, **248,** 2645–2649 (1973).
75. S. Siegel, A. H. Brady, and W. M. Awad, Jr., *J. Biol. Chem.*, **247,** 4155–4159 (1972).
76. Y. D. Halsey and H. Neurath, *J. Biol. Chem.*, **217,** 247–252 (1955).
77. R. L. Heinrikson, *Methods Enzymol.*, **42,** 175–189 (1977).
78. P. Cupo, W. El-Deiry, P. L. Whitney, and W. M. Awad. Jr., *J. Biol. Chem.*, **255,** 10,828–10,833 (1980).
79. V. P. Torchilin, A. V. Maksimenko, V. N. Smirnov, I. V. Berezin, A. M. Klibanov, and K. Martinek, *Biochim. Biophys. Acta,* **522,** 277–283 (1978).
80. A. M. Klibanov, *Anal. Biochem.*, **93,** 1–25 (1979).
81. J. J. Marshall, *Trends Biochem. Sci.*, **3,** 79–83 (1978).
82. P. Tuengler and G. Pfleiderer, *Biochim. Biophys. Acta,* **484,** 1–8 (1977).
83. H. Nishimura, A. Matsushima, and Y. Inada, *Enzyme* (*Basel*), **26,** 49–53 (1981).
84. F. Macritchie, *Adv. Protein Chem.*, **32,** 283–326 (1978).
85. P. L. Felgner and J. E. Wilson, *Anal. Biochem.*, **74,** 631–635 (1976).
86. K. J. Kramer, P. E. Dunn, R. C. Peterson, H. L. Seballos, L. L. Sandburg, and J. H. Law, *J. Biol. Chem.*, **251,** 4979–4985 (1976).

87. R. W. Blakesley and J. A. Boezi, *Biochim. Biophys. Acta,* **414,** 133–145 (1975).
88. H. Tenenbaum-Bayer and A. Levitzki, *Biochim. Biophys. Acta,* **445,** 261–279 (1976).
89. J. A. Smith, J. G. R. Hurrell, and S. J. Leach, *Anal. Biochem.,* **87,** 299–305 (1978).
90. C. H. Suelter and M. DeLuca, *Anal. Biochem.,* **135,** 112–119 (1983).
91. M. Dixon and E. C. Webb, *Enzymes,* 3rd ed., Academic, New York, 1979, p. 11.
92. C. R. Thomas and P. Dunnill, *Biotechnol. Bioeng.,* **21,** 2279–2302 (1979).
93. S. E. Charm and B. L. Wong, *Enzyme Microb. Technol.,* **3,** 111–118 (1981).
94. P. Bartholmes, H. Durchschlag, and R. Jaenicke, *Eur. J. Biochem.,* **39,** 101–108 (1973).
95. P. D. Boyer, *The Enzymes,* 2nd ed., Vol. 1, Academic, New York, 1959, pp. 511–588.
96. R. Stevens, L. Stevens, and N. C. Price, *Biochem. Education,* **11,** 70 (1983).
97. G. G. Hawley, *The Condensed Chemical Dictionary,* 9th ed., van Nostrand–Reinhold, New York, 1977.
98. W. W. Cleland, *Biochemistry,* **3,** 480–482 (1964).
99. M. Friedman, *The Chemistry and Biochemistry of the Sulfhydryl Group in Amino Acids, Peptides, and Proteins,* Pergamon, Oxford, 1973, pp. 1–87.
100. U. T. Ruegg and J. Rudinger, *Methods Enzymol.,* **47,** 111–116 (1977).
101. C. H. Suelter and E. E. Snell, *J. Biol. Chem.,* **252,** 1852–1857 (1977).
102. T. L. Coombs, J. P. Felber, and B. L. Vallee, *Biochemistry,* **1,** 899–905 (1962).
103. C. L. Zielke and C. H. Suelter, *J. Biol. Chem.,* **246,** 2179–2186 (1971).
104. A. Holmgren, *Curr. Top. Cell. Regul.,* **19,** 47–76 (1981).
105. C. B. Anfinsen, *Biochem. J.,* **128,** 737–749 (1972).
106. R. Jaenicke and R. Rudolph, Folding and association of oligomeric enzymes, in R. Jaenicke, Ed., *Proceedings of the 28th Conference German Biochemical Society,* Elsevier–North Holland, Amsterdam, 1979, pp. 525–548.
107. E. W. Westhead, *Biochemistry,* **3,** 1062–1068 (1964).
108. R. Rudolph, G. Zettlmeissl, and R. Jaenicke, *Biochemistry,* **18,** 5572–5575 (1979).
109. M. Tobes, R. T. Kuczenski, and C. H. Suelter, *Arch. Biochem. Biophys.,* **151,** 56–61 (1972).
110. D. H. Porter and J. M. Cardenas, *Biochemistry,* **19,** 3447–3452 (1980).
111. K-P. Wong and L. M. Hamlin, *Arch. Biochem. Biophys.,* **170,** 12–22 (1975).
112. G. M. Stancel and W. C. Deal, Jr., *Biochemistry,* **8,** 4005–4011 (1969).
113. K. Weber and D. J. Kuter, *J. Biol. Chem.,* **246,** 4504–4509 (1971).
114. S. N. Vinogradov and O. H. Kapp, *Methods Enzymol.* **91,** 259–263 (1983).
115. W. H. Konigsberg and L. Henderson, *Methods Enzymol.* **91,** 254–259 (1983).
116. T-H. Liao, *J. Biol. Chem.,* **250,** 3831–3836 (1975).
117. S. A. Lacks and S. S. Springhorn, *J. Biol. Chem.,* **255,** 7467–7473 (1980).
118. A. Spanos and U. Hubscher, *Methods Enzymol.,* **91,** 263–277 (1983).
119. A. Blank, J. R. Silber, M. P. Thelen, and C. A. Dekker, *Anal. Biochem.,* **135,** 423–430 (1983).
120. D. A. Hager and R. R. Burgess, *Anal. Biochem.,* **109,** 76–86 (1980).
121. S. Clarke, *Biochim. Biophys. Acta,* **670,** 195–202 (1981).
122. M. J. Heeb and O. Gabriel, *Methods Enzymol.,* **104,** 416–439 (1984).

2

Protein Concentration and Enzyme Activities

Before discussing the purification of an enzyme (Chapter 3), it is important to understand the problems involved in determining the amount of protein and the units of enzyme activity in a sample. Not only are these data required for the development of an enzyme purification scheme, but protein concentrations and enzyme activities are also routinely needed for research, for the diagnosis of diseases, and for many other purposes. For those reasons, a review of methods for determining protein concentrations and enzyme catalytic activities are presented in this chapter.

DETERMINING PROTEIN CONCENTRATIONS

Many methods of differing sensitivities and procedural complexities have been developed over the last 75 years to determine protein concentration. Most are based on (a) the intrinsic property of proteins to absorb ultraviolet (UV) light, (b) the formation of chemical derivatives, or (c) the ability of proteins to bind dyes (1). The most useful methods and their sensitivities are outlined in Table 2.1; advantages and disadvantages of each are described.

Absorption Methods

The spectrophotometric determination of protein is rapid, sensitive, and the protein sample is not destroyed by the determination. It is the method of choice when using a pure protein with a known extinction coefficient

Table 2.1 Sensitivity range of several methods useful for determining protein concentrations

Method	Sensitivity Range (μg)	Extinction Coefficient or Concentration Equation	Reference
Absorption methods			
A_{280}[a]	100–3000	$\epsilon_{280} = 1$ mL/mg cm	2
A_{205}[b]	3–100	$\epsilon_{205} = 31$ mL/mg cm	3
$A_{280} - A_{260}$[c]	100–3000	Protein (mg/mL) = $(1.55A_{280} - 0.76A_{260})$	4
$A_{235} - A_{280}$[d]	25–700	Protein (mg/mL) = $(A_{235} - A_{280})/2.51$	5
$A_{224} - A_{236}$	5–180	*e*	6
$A_{215} - A_{225}$	2–45	Protein (μg/mL) = $144(A_{215} - A_{225})$	7

[a compilation of the extinction coefficients of many proteins is given in Reference (12)]. Measurements at 280 nm are useful and generally adequate to estimate protein concentrations in crude homogenates and partially purified samples during the development of a protein purification scheme. When samples of crude homogenate contain suspensions of insoluble protein, they can often be clarified by adding sodium dodecyl sulfate [0.1% w/v final concentration (13)] before measuring absorbances. If the sample contains sulfhydryl reducing agents that absorb strongly in the UV region, they must be removed prior to measuring the absorbances (procedures for removing them are discussed later in this chapter). Because pure proteins have widely different extinction coefficients, measurement of protein concentration in samples during the later stages of a purification should be assessed by more general methods such as the Lowry method (14).

Microbial extracts usually contain significant amounts of nucleic acid that absorb at 280 nm, so that the quantification of protein in these extracts based solely on measurements at 280 nm will be in serious error. To estimate the contribution of contaminating nucleic acids to the 280-nm absorption, measure absorbances of the same sample at 260 nm as suggested by Warburg and Christian (11). Protein concentrations can then be calculated using Equation (2.1) (3).

$$\text{Protein (mg/mL)} = (1.55A_{280} - 0.76A_{260}) \tag{2.1}$$

Table 2.1 (*Continued*)

Method	Sensitivity Range (μg)	Extinction Coefficient or Concentration Equation	Reference
Derivative methods			
Lowry	25–100 at 500 nm[f] 2–30 at 660 nm 1–20 at 750 nm	Use standard curve	8
Biuret	1000–10,000	ϵ_{545} = 0.06 mL/mg cm	2
Fluorescamine	1–50 Excitation wavelength at 390 nm Emission wavelength at 475 nm	Use standard curve	9
o-Phthalaldehyde	1–5 Excitation wavelength at 340 nm Emission wavelength at 475 nm	Use standard curve	1
Dye binding method (the Bradford assay)			
Coomassie Brilliant Blue G-250	1–15	ϵ_{595} = 81 mL/mg cm[g]	10

[a] For bovine serum albumin, ϵ_{280} = 0.70 mL/mg cm (5).

[b] To minimize stray light effects, limit absorption at 205 nm to 0.7*A*. For bovine serum albumin, ϵ_{205} = 29.6 mL/mg cm (3).

[c] Layne (4) developed this equation with the data provided by Warburg and Christian (11) to correct for nucleic acid contamination.

[d] Bovine serum albumin at 1 mg/mL gives $(A_{280} - A_{235})/2.51 = 0.94$ (5).

[e] Bovine serum albumin at 1 mg/mL gives $(A_{224} - A_{236}) = 0.6$ (6).

[f] These concentration ranges give linear standard curves in the modified Lowry method (see Inset 2.1). They represent the total amount of protein in a final assay volume of 1.3 to 1.4 mL. The range can be extended by working in the nonlinear portion of the standard curve. Log–log plots or hyperbolic functions also extend the range of the working standard curve (1).

[g] This value is an approximation since the standard curve is not linear, particularly at low protein concentrations.

This equation is obtained from absorbances at 260 and 280 nm of artificial mixtures of yeast enolase and nucleic acids (11).

Whitaker and Granum (5) and Groves et al. (6) describe more accurate spectrophotometric methods for determining protein concentration, particularly for samples containing significant amounts of nucleic acid. Whitaker and Granum note that the differences between the absorbances at 280 and 235 nm for a protein is due primarily to the peptide bond. Furthermore, nucleic acids have essentially the same absorbance values at these two wavelengths. As a consequence, Whitaker and Granum's method is essentially independent of the specific amino acid composition of the protein and is not severely affected by nucleic acid contamination. Equation (2.2) can then be used to calculate protein concentrations from measurements at 235 and 280 nm.

$$\text{protein (mg/mL)} = (A_{235} - A_{280})/2.51 \qquad (2.2)$$

Absorbance measurements at these wavelengths can usually be made on the same dilution since the average measured extinction coefficient for eight proteins at 235 nm is 2.8 times the 280-nm value.

The method of Groves et al. (6) is also useful when protein samples contain relatively large amounts of nucleic acid. In this case, protein concentrations are calculated from differences in absorbance at 224 nm and an isoabsorbance wavelength of the contaminating nucleic acids near 236 nm. The accuracy of the determination depends on how well the isoabsorbance wavelength is defined. An accurate determination of this wavelength requires isolation of the contaminating nucleic acids. However, the errors due to an improper isoabsorbance wavelength are negligible at low nucleic acid contamination. By using 236 nm, satisfactory results can be obtained.

Another spectrophotometric method for determining protein concentrations was recently reevaluated by Wolf (7). This method is based on differences in absorbance between 215 and 225 nm. The method gives the best results when several measurements on serially diluted samples are made. Protein concentrations are then calculated with Equation (2.3)

$$\text{protein } (\mu\text{g/mL}) = 144(A_{215} - A_{225}) \qquad (2.3)$$

The recent results indicate that the optimum range of protein concentrations for this method varies from 2 to 45 μg/mL. There is no evidence to indicate that contaminating nucleic acids interfere with this assay.

Absorbance measurements at short wavelengths (between 200 and 240 nm) as required by the methods of Whitaker and Granum (5), Groves et al. (6), and others (7), are often in error primarily due to the effects of scattered light. To monitor the extent of this error, use bovine serum albumin as a standard; Table 2.1 gives its extinction coefficient for each of the absorption methods.

Lowry Method

The most widely used method for determining protein was published by Lowry et al. in 1951 (14). This paper was cited in over 10,000 journal articles in 1981 alone (15). The method is simple and sensitive (see Inset 2.1), and many proteins give the same color response on a milligram per milliliter basis. The principal disadvantage in using the Lowry method is the interference caused by a large number of substances summarized in Table 2.2, which was compiled by Peterson (16).

The Lowry method is based on the reaction of certain amino acids in a protein and a mixture of sodium molybdate and sodium tungstate in phosphoric acid; the active constituents of the Folin–Ciocalteu phenol reagent (16). Tyrosine, tryptophan, and to a lesser extent, cystine, cysteine, and histidine, are the principal amino acids that react. The amino acids reduce the mixed metalloacids by causing loss of one to three oxygen atoms from tungstate and/or molybdate, thereby producing one or more of several possible compounds with a characteristic blue color. The electron transfer to the mixed metalloacid is apparently facilitated by the chelation of copper to the peptide backbone. Peptide linkages also affect reduction of the mixed acids because complete hydrolysis of a protein reduces its color yield by one half to two thirds (16).

Most substances interfere by causing an increase in the absorbance of the reagent blank, although some substances also interfere with the color development (1, 16). Nonionic and cationic detergents, lipids, and some salts cause precipitates that interfere with the spectrophotometric measurements. Most interferences can be eliminated by incorporating the substance into the reagent blank and the protein standard. Interference by some substances can be eliminated by dialysis, by evaporation, or by chemical modification of the interfering substance. Sometimes it may be necessary to remove the interfering compound by precipitating the protein and then determining the amount of protein in the precipitate.

INSET 2.1

Reagents and Procedures for Determining the Concentration of Proteins by Derivative Methods

---

Standard Lowry (14)

Reagent A: 2% Na_2CO_3 in 0.1 *N* NaOH

Reagent B: 0.5% $CuSO_4 \cdot 5\ H_2O$ in 1% sodium tartrate or 1% trisodium citrate (2)

Reagent C: Mix 50 mL reagent A with 1 mL reagent B; discard after 1 day

Reagent D: Dilute commerical Folin–Ciocalteu reagent with water to 1 *N* in acid, an approximate 1:1 dilution. This diluted reagent is as stable as the original stock (16). Discard commerical reagent if $A_{660\ nm} > 0.2$; alternatively, add bromine to reoxidize (17).

Modified Lowry (8)

Reagent A: 2% Na_2CO_3, 1% sodium dodecyl sulfate (SDS), 0.16% sodium tartrate in 0.1 *N* NaOH

Reagent B: 4% $CuSO_4 \cdot 5H_2O$ in water

Reagent C: Mix 100 mL reagent A with 1 mL reagent B; discard after 1 day

Reagent D: Same as reagent D in standard Lowry

Comment

The modified Lowry allows assay of protein samples containing up to 2.5 m*M* ethylenediamintetraacetic acid (EDTA), 200 m*M* sucrose, lipoproteins, Triton X-100, and other detergents. It is important to realize that this modification enhances the sensitivity to interference by Tris (8).

Procedure

To a 0.2- or 0.3-mL protein sample, add 1 mL of reagent C and allow to stand at room temperature for at least 10 min. A longer time (up to 24 hr) will not create a problem unless the sample contains hydrolyzable lipids.

Add 0.1 mL reagent D while mixing rapidly. Let stand for 30 min or longer (up to 2 hr) and determine absorption at 500, 660, or 750 nm depending on the sensitivity required. Sodium dodecyl sulfate may be omitted from reagent A of the modified Lowry and added to the reaction mix before measuring the absorbance (8). The reaction volumes may be scaled up or down to either increase or decrease the amount of protein required for an assay.

Bovine serum albumin is used most often as a standard. Determine its concentration spectrophotometrically, $\epsilon_{280\ nm}$ = 0.70 mL/mg cm.

Biuret (17)

Reagent

Dissolve 3.8 g $CuSO_4 \cdot 5H_2O$ and 6.7 g NaEDTA in 700 mL H_2O. While stirring, slowly add 200 mL 5 *N* NaOH (18) and then 1 g KI to prevent deposition of Cu_2O. Store in a plastic container.

Procedure

Add 1 mL of the biuret reagent to a 0.1-mL protein sample (2–24 mg/mL); let stand 25 min and read at 545 nm. The color is stable for 1 hr. Use bovine serum albumin as a standard.

INSET 2.1 (*Continued*)

Dye binding method (10) (the Bradford assay)

Reagent

Combine 10 mg Coomassie Brilliant Blue G-250 (Color Index 42,655) with 10 mL 88% phosphoric acid and 4.7 mL absolute ethanol, and dilute to 100 mL with distilled water. Since dyes from different sources vary in dye content, the $A_{550\ \mathrm{nm}}$ should be 1.18. If it is not, add additional dye, filter solution through Whatman No. 1 paper, and store in dark colored bottles. This dye solution is near saturation so precipitates may appear with time and filtration and recalibration may be required. Shelf life is several weeks (10). Of several dye sources tested, Serva Blue G from Serva Feinbiochemica is the purest. It is chemically identical to Coomassie Brilliant Blue G-250.

Procedure

Add 950 μL dye reagent to a 50-μL protein sample containing 0.1 to 5.0 μg protein. Read $A_{595\ \mathrm{nm}}$ after at least 4 min have elapsed against a blank of 50 μL of buffer added to the 950 μL of dye reagent. The color is stable for 1 hr.

Table 2.2 Tolerable concentration limits of substances tested directly for interference in the Folin phenol (Lowry) protein quantitation method[a]

Compound Tested[b]	Tolerable Concentration Limit in Final Sample Volume[c]
Amine derivatives (except those otherwise cataloged below)	
Trimethylamine, methylethanolamine	2.5 μ*M*
Triethylamine, tripropylamine, tributylamine, dimethylglycine, dimethylethanolamine, triethanolamine, diethanolamine	750 n*M*
Methylamine, ethanolamine, dimethylamine, sarcosine, betaine, carnitine, trimethylamine, sarcosine, betaine, carnine, trimethylamine-*N*-oxide	>82 m*M*
Putrescine, spermine, tetramethylammonium bromide, choline	<42 m*M*
Iminodiacetic acid, EDTA	<200 μ*M*
Spermidine	<8.2 m*M*
Histamine	5 μ*M*
Amino acids	
Glycine	2.5 m*M*
Tyrosine, tryptophan, 5-hydroxytryptophan	380 n*M*
Cysteine, cystine	100 μ*M*
Histidine	38 μ*M*
Buffers[d] (pH in standard buffering range)	
Tris, Tricine	250 μ*M*
Bicine	12 μ*M*
Glycylglycine	250 μ*M*
Sodium citrate	2.5 m*M*
Sodium phosphate (pH 8.9)	250 m*M*
Hepes, Hepps	2.5 μ*M*
Pipes, Bes, Hida	5 μ*M*
Mops, Mes, Caps, Ada, Aces, Taps	25 μ*M*
Tes, Ches	1.0 m*M*
Chelating agents[d]	
EDTA	125 μ*M*
GEDTA	7.5 μ*M*
Detergents	
Triton X-100	2.5 mg/mL
Triton X-155, Tween 60, Terric 200, Hyamine 2389	<1.25 mg/mL
Triton WR-1339	25 μg/mL
Sodium deoxycholate	625 μg/mL
SDS	12.5 mg/mL

Table 2.2 (*Continued*)

Compound Tested[b]	Tolerable Concentration Limit in Final Sample Volume[c]
Drugs	
Sodium salicylate, acetylsalicylic acid, terramycin, *p*-aminosalicylic acid	50 ng/mL
Chlorpromazine, sulfamerazine, sulfadiazine, sulfanilamide, streptomycin	750 ng/mL
Phenacetin, sulfaguanidine, chloramphenicol	1.75 µg/mL
Erythromycin, sodium Veronal	>200 µg/mL
Penicillin, sodium doxacillin monohydrate	12.5 units/mL
Methotrexate	250 µ*M*
Sodium aminohippurate	200 ng/mL
Catecholamines (from adrenal gland)	?
Hexosamines	
Glucosamine, galactosamine	12.5 µg/mL
Mannosamine	>115 µg/mL
N-Acetyl-D-glucosamine, *N*-acetyl-D-galactosamine	250 µg/mL
Lipids, fatty acids, and so on	
Lipids	?
Fatty acids oxidized in 0.05 *N* NaOH, 37°C, 18 hr	
Oleic acid	500 µ*M*
Arachidonic acid	100 µ*M*
Phosphatidylethanolamine, cardiolipin, cholesterol, phosphatidylcholine, sulfatide, phosphatidylinositol	25 µ*M*
Fatty acids not oxidized (those above)	2.5 m*M*
Sphingomyelin, triolein, palmitic acid, cerebroside, ethanolamine, serine	125 µ*M*
Miscellaneous compounds and reducing agents	
Ethylene glycol	2.5 µL/mL
Polyvinyl pyrrolidone	≤62 µL/mL
Dimethyl sulfoxide	≥62 µL/mL
Urea	>200 m*M*
Succinic acid	10 m*M*
Perchloric acid (neutralized), trichloroacetic acid (neutralized)	≥12.5 mg/mL
Ethanol, ether	>125 µL/mL
Acetone	>12.5 µL/mL
Acetylacetone	5 µ*M*
Elemental sulfur	?
Ferrichrome, acetohydroxamic acid, benzohydroxamic acid	75 ng/mL
Hydrazine, hydroxylamine	>20 µg/mL
Hematin	38 ng/mL

Table 2.2 (*Continued*)

Compound Tested[b]	Tolerable Concentration Limit in Final Sample Volume[c]
Polyacrylamide gel electrophoresis materials	
Extract of 8.3% polyacrylamide gel	?
Acrylamide	1.25 mg/mL
Potassium ferrocyanide	5 μg/mL
Ammonium persulfate	>1 mg/mL
Tris plus Temed[d]	<(2.8 + 0.05)μg/mL
Potassium ferricyanide, ferric chloride, ammonium sulfamate, sodium bisulfite, sodium thiosulfate	>3.8 mg/mL
Dithioerythritol, oxidized glutathione	50 μ*M*
Reduced glutathione	100 μ*M*
2-Mercaptoethanol	1.75 m*M*
Sodium dithionite, *d*-aminolevulinic acid, levulinic acid, ascorbic acid, stannous chloride, ferrous chloride	250 ng/mL
Acetylaldehyde, furfural, 5-hydroxymethylfurfural	38 μg/mL
Nucleic acids, nucleotides, and so on	
DNA (in 1 *M* NaCl)	190 μg/mL
Guanine, xanthine	250 m*M*
Guanylic acid	1.25 μ*M*
Hypoxanthine, guanosine, deoxyguanosine, uracil	7.5 μ*M*
Deoxyguanylic acid, adenosine, cytidine	50 μ*M*
Adenine, cytosine, thymine, uridine, thymidine, adenylic, cytidylic, uridylic, thymidylic, orotic acids	1.9 m*M*
Guanidine	>210 m*M*
Salts	
Sodium chloride	1.75 *M*
Sodium bromide	>750 m*M*
Sodium iodide	38 m*M*
Sodium tungstate, barium hydroxide	>42 m*M*
Sodium nitrate	>300 m*M*
Sodium sulfate	>175 m*M*
Sodium phosphate	>100 m*M*
Potassium phosphate, potassium chloride	30 m*M*
Potassium iodide	38 m*M*
Potassium cyanide	80 μ*M*
Potassium cyanate, potassium thiocyanate	>800 μ*M*
Ammonium sulfate	>28 m*M*
Zinc sulfate	>15 m*M*
Cobalt chloride	2.5 μ*M*
Manganese chloride, mercury chloride	38 μ*M*
Magnesium ions	?

Table 2.2 (*Continued*)

Compound Tested[b]	Tolerable Concentration Limit in Final Sample Volume[c]
Sugars and their polymers	
Sucrose	10 mM
Fructose, glucose, sorbose, xylose, rhamnose, mannose, galactose	>30 mM
Tagatose	85 μM
Ribose	>5 mM
Glycerol	250 μL/mL
Inulin, glycogen, trehalose, dextran	>775 μg/mL
Dextran	1.25 mg/mL
Metrizamide	25 mg/mL
Ficoll	75 μg/mL
Chondroitin-6-sulfate	>3.75 mg/mL
Heparin	>125 μM

[a] Adapted by permission from Reference (1).
[b] The compounds tested are grouped under general categories, and the categories are arranged alphabetically. Compounds fitting into more than one category are listed only once and appear in the first appropriate category.
[c] The tolerable concentration limit is that at which there is no interference in the protein assay. Lack of interference is defined as < 2% error in protein estimate and < 0.002 absorbance unit for the reagent blank. In cases where the effect on the color yield due to protein was not studied, the tolerable limit is determined from the effect on the reagent blank alone. In cases where complete information was unavailable, the tolerable limit was established, assuming the effect on the reagent blank or the protein estimate was linear with respect to concentration of the interfering substance. In such cases, any error in the estimate of the tolerable limit would usually be an underestimate, as judged from the behavior of the majority of those interfering substances that do show nonlinear effects. Figures preceded by a less than (<) or greater than (>) symbol indicate that the tolerable limit is unknown but is, respectively, less than or greater than the figure shown.
[d] Abbreviations: Hepes, *N*-2-hydroxyethylpiperazine-*N'*-2-ethanesulfonic acid; Hepps, *N*-2-hydroxyethylpiperazine-*N'*-3-propanesulfonic acid; Pipes, piperazine-*N*,*N'*-bis(2-ethanesulfonic acid); Bes, *N*,*N*-bis(2-hydroxyethyl)-2-aminoethanesulfonic acid; Hida, 2-hydroxyethyliminodiacetic acid; Mops, 3-(*N*-morpholino)propanesulfonic acid; Mes, 2-(*N*-morpholino)ethanesulfonic acid; Caps, 3-(cyclohexylamino)-1-propanesulfonic acid; Ada, *N*-(2-acetamido)-2-iminodiacetic acid; Aces, *N*-(2-acetamido)-2-aminoethanesulfonic acid; Taps, N-[tris(hydroxymethyl)methyl]-3-aminopropanesulfonic acid; Tes, *N*-tris(hydroxymethyl)methyl-2-aminoethanesulfonic acid; Ches, 2-(cyclohexylamino)ethanesulfonic acid; Tris, tris(hydroxymethyl)aminomethane; Bicine, *N*,*N*-bis(2-hydroxyethyl)glycine; Tricine, *N*-tris(hydroxymethyl)methylglycine; Gedta, glycolethylenediamine-*N*,*N*,*N'*,*N'*-tetraacetic acid; Temed, *N*,*N*,*N'*,*N'*-tetramethylethylenediamine

If the protein is first precipitated by acid, collect the precipitate by centrifugation. Adding sodium deoxycholate (19, 20) or yeast soluble RNA (21) (both at 125 μg/mL final concentration) as a coprecipitating carrier before adding trichloroacetic acid gives reproducible and essentially quantitative precipitation of 5 to 50 μg/mL bovine serum albumin. If the sample contains detergent, add ribonucleic acid because deoxycholate does not yield a precipitate with trichloroacetic acid in the presence of detergent (21).

Sulfhydryl compounds at 1–2 m*M* are removed by oxidation with H_2O_2 (22), ammonium persulfate (23), or by reaction with *N*-ethylmaleimide, with little effect on the sensitivity (24). Because 2-mercaptoethanol is more volatile, it may be removed by heating or preferably by vacuum drying in a boiling H_2O bath (25).

Precipitates that form during the Lowry assay can often be removed from the final reaction by centrifugation. Interferences caused by lipids or nonionic detergents can be eliminated by adding sodium dodecyl sulfate (1% final), either after the reaction is completed, or to the reagents before initiating the reaction (8). Using the modified version of the Lowry procedure (8) (see Inset 2.1) will not only reduce interference by lipids and detergents but also by EDTA and sucrose.

The standard curve of optical density versus protein concentration obtained by the Lowry method is not linear, although portions of the curve approximate linearity. As indicated in Table 2.1, measurements made at different wavelengths have differing extinction coefficients so that limited linear standard curves can be obtained over three different concentration ranges. Linear standard curves can be obtained over large concentration ranges by a log–log plot (1) or a double reciprocal plot (26). The equation for the double reciprocal plot has the form

$$1/A = m/[\mathrm{P}] + b$$

where A is absorbance, [P] the protein concentration, m the slope of the line, and b the intercept on the $1/A$ axis. The values of m and b can be obtained directly from the double reciprocal plot or by using a linear regression program. For routine analysis, one may calculate the values of m and b from the equations

$$m = (1/A_{\mathrm{l}} - 1/A_{\mathrm{h}})/(1/[\mathrm{P_l}] - 1/[\mathrm{P_h}])$$
$$b = 1/\bar{A}_{\mathrm{l}} - \mathrm{m}/[\bar{\mathrm{P}}]$$

where A_l and A_h are the absorbances for two protein samples of widely separated low and high concentrations $[P_l]$ and $[P_h]$. $\bar{A}$ and $[\bar{P}]$ are the means of the absorbances $(A_l + A_h)/2$, and protein concentrations $([P_l] + [P_h])/2$ as noted in the preceding equation. The protein concentration of each unknown is then calculated from the equation

$$[P] = mA/(1 - bA)$$

As with all standard curves, but particularly with the Lowry method, it is important to show that the absorbances of the blank are correctly measured.

Biuret Method

Although the biuret method is about 100 times less sensitive than the Lowry procedure, it does offer some advantages. There is little variation in the color intensity given by different proteins since the color is due to a complex between peptide bonds and copper, $\epsilon_{540} = 0.06$ mL/mg cm. Relatively few substances interfere; of those that do, ammonium ions and lipids cause the most difficulty. Ammonium ions form cuproammonium compounds, therefore negating the use of the biuret method on protein fractions containing ammonium sulfate (2). Lipids cause turbidity, however, the turbidity usually develops more slowly than the color reaction, so the color may be read between 5 and 15 min (27) instead of the normal 25 min. Turbidity can be eliminated by adding sodium deoxycholate (final concentration 3% w/v) to the protein sample. When the solution clears—in about 10 min—add the biuret reagent (2). Reproducible results can also be obtained by reading the color in 10 sec with a stopped-flow device (28). If a biuret reaction fails to give sufficient color for accurate measurement because of insufficient protein, the Lowry method may be applied to the same reaction solution (29).

Dye Binding Method (Bradford Assay)

Despite the large variation of the dye binding assay (30) in response to different proteins (31, 32), the method is still popular: Authors of over 1200 journal articles cited this paper in 1981 (15). The Coomassie Brilliant Blue G-250 or the Serva Blue G dye used in the dye binding method exists in acid solution in two forms, a blue form and an orange form. Proteins preferentially bind the blue form to produce a complex with an extinction coefficient much greater than the free dye (10).

This assay is sensitive, simple, quick, inexpensive, and very few substances interfere with the determination. Only detergents, alkaline protein solutions, and Ampholines interfere so much that the reagent blank cannot be used to correct it (10). Sodium dodecyl sulfate can be precipitated with KCl (33) and alkaline protein solutions can be neutralized before performing the assay (10). If desired, the amount of protein in a sample can also be determined by precipitating the protein–dye complex and then determining the amount of dye in the precipitate (34, 35).

A recent modification of the dye binding assay appears to remove much of the variation in the original method (10). Some of the variation reported previously is due to the differences in the purity of the dye obtained from various commercial sources. Even though the actual amount of dye in samples from commercial sources varies by as much as fivefold, dyes from different sources give the same protein color yield if the concentrations of each are the same. If the contaminants in a dye sample are not soluble, filter the dye reagent before use. It must be emphasized that even if the dye concentrations are properly adjusted, highly cross-linked proteins such as ribonuclease, chymotrypsin, and trypsin, and proteins with a low pI, will still give a poor color response (31, 32).

Fluorescence Methods

Fluorescamine and *o*-phthalaldehyde react with primary amines to form fluorescent derivatives. The fluorescamine reagent is prepared in dry dioxane, acetonitrile, or acetone and added to the protein solution while mixing vigorously with a vortex mixer (9). Rapid mixing is essential because the reagent is readily hydrolyzed by water. At pH 8–9, the reaction with amines is complete in seconds and the excess reagent reacts with water within about 1 min to yield nonfluorescent derivatives. Secondary amines and ammonium ions consume reagent but do not yield fluorescent products. Alcohols, on the other hand, form reversible complexes with fluorescamine effectively slowing both the hydrolysis reaction and the reaction with amines (9). The fluorescence of the derivatized protein is measured at 475 nm while using 390 nm exciting light.

o-Phthalaldehyde also reacts rapidly with primary amines to form fluorescent products. After the reaction is complete, in about 15 min, NaOH (0.5 *M* final concentration) is added to stabilize the fluorescent reaction product and destroy most of the fluorescence in the blank due to ammonia. The fluorescence is measured at 450 nm while exciting at 340 nm.

Two principal problems with fluorescent assays are (a) many protein samples contain amine contaminants that give fluorescent derivatives and (b) not all proteins exhibit the same fluorescence yield. Contaminating small molecular weight amines may be removed by collecting the fluorescent labeled proteins quantitatively on nitrocellulose membrane filters that allow the labeled small molecular weight compounds to pass. After filtration, the filter containing the fluorescamine labeled protein is placed in a solution of 0.2 *M* sodium borate, pH 9.0, and acetone (2.5:1, v/v) and gently mixed. Measure the flourescence after 10 min (36).

Because different proteins exhibit different fluorescent yields, the protein selected as a standard will affect the final result. Some investigators use the same protein as that being measured as a standard, others select an arbitrary standard such as bovine serum albumin, or subject the protein to acid hydrolysis before completing the fluorescent assay (9). Most investigators use the arbitrary standard.

If the protein is to be hydrolyzed before applying the fluorescent method, use 6 *N* HCl for 4–24 hr. The protein concentration must be diluted (below 20 μg/mL) to insure reliability (1). After subjecting a protein sample to hydrolysis, the free amino acids are analyzed by either of the fluorescent methods. To calculate the protein in the sample, multiply the total moles of amino acids released by a conversion factor, which is the average molecular weight of the specific amino acids in the protein corrected for loss of tryptophan (1). The average molecular weight of the amino acids in most proteins falls between 110 and 120. For bovine serum albumin, the number is 121. For 184 enzymes that have been sequenced (37), the average molecular weight of the amino acids is 112. The average molecular weight of the amino acids in the proteins listed in Table 4.4 is 111 ± 4 (one standard deviation). Correcting for the loss of tryptophan during acid hydrolysis, the lack of reactivity of proline in the fluorescent analysis, and the different fluorescent yields of each amino acid relative to the standard alanine gives 127.4 for the conversion factor (1).

The fluorescent reagent OPA is preferred over fluorescamine for determining protein concentration because it is less expensive, less pH sensitive, has a greater sensitivity to amino acids, and is soluble and stable in aqueous buffers.

Miscellaneous Methods

Other methods utilizing a variety of substances and reactions such as ninhydrin (38), radioactive isotopes (39), and turbidity (40, 41) are sometimes used to assay proteins when the procedures discussed above are not applicable. Refer to Peterson (16) for a brief discussion of some of these miscellaneous methods. A recent report (42) shows that microgram quantities of protein can be analyzed by the spectrophotometric measurement of trinitrobenzene sulfonate derivatives of amino acids in hydrolyzates. Amino acids in protein hydrolyzates react with 2,4,6-trinitrobenzene sulfonic acid to give derivatives that have an average molar absorbance of $1.9 \times 10^4/(M\text{cm})^{-1}$ at 416 nm. The hydrolyzate of BSA reacts with the trinitrobenzene reagent to give ϵ_{416} = 150 mL/mg cm.

EVALUATION OF PROTEIN QUANTITATION METHODS

The sensitivities of various protein quantitation methods are compared in Figure 2.1, and their responses with different proteins are given in Table 2.3. Fluorescent methods are the most sensitive; if the protein is first hydrolyzed, as little as 10 ng can be quantitated. Hydrolysis also eliminates much of the variability between the fluorescence yields of different proteins. The Lowry, biuret, and the dye binding methods are relatively simple and, except for the biuret procedure, also relatively sensitive. The principal disadvantage of these methods is that many substances interfere, not only in the reagent blank, but also in formation of the chromophore itself.

Measuring protein concentration by absorbance methods has an advantage in that the sample is not destroyed during the measurement. Furthermore, with most of the absorbance methods, interference by nucleic acids can be eliminated. The principal disadvantage is due to nonprotein substances that absorb or scatter light particularly in the 205–240-nm region.

MEASUREMENT OF ENZYME ACTIVITIES

Initial Velocity

The amount of enzyme in a sample is almost always expressed in terms of the rate (micromoles per minute) at which an aliquot of the sample

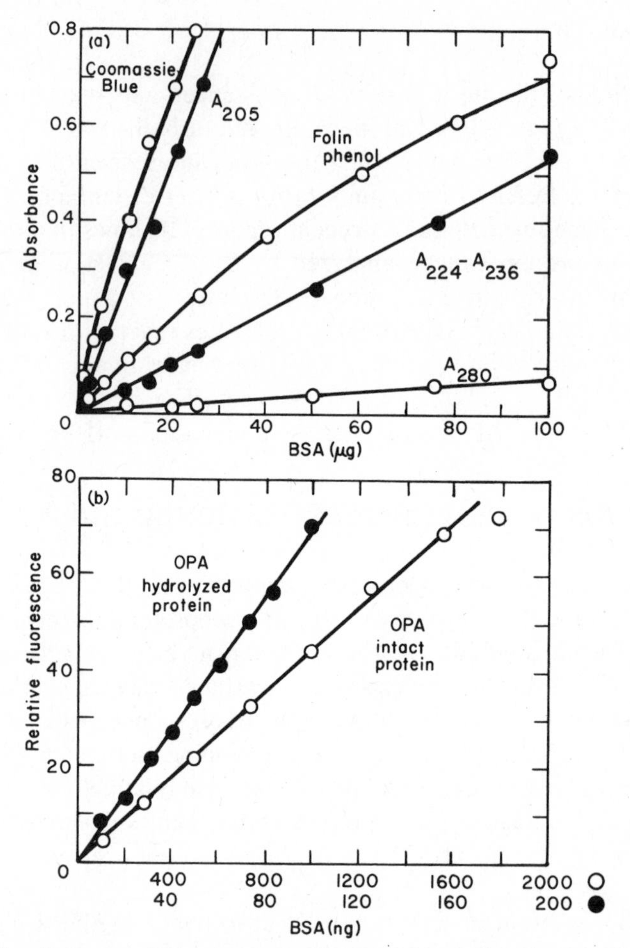

Fig. 2.1. Standard curves for bovine serum albumin (BSA) obtained by the various methods of protein quantitation. (*a*) Spectrophotometric procedures. The quantity of BSA shown is the amount present in a 1.0-mL sample volume for the standard assay procedures. The nonlinear functions for the Folin phenol and Coomassie Brilliant Blue G-250 standard curves were fitted according to a hyperbolic function [Equations (1), (2), and (5)–(7) of Reference (1)] for the former and a power function [Equations (3), (4), (8), and (9) of Reference (1)] for the latter. (*b*) Fluorometric procedures. The quantity of BSA shown is the amount present in the 10–μL sample volume of the standard assay. Note the different scales used between the spectrophotometric and flurorometric assays and between the hydrolyzed and intact protein assays of the *o*-phthalaldehyde (OPA) fluorometric procedures. Abstracted with permission from Reference (1).

Table 2.3 Comparison of the relative responses with different proteins analyzed by the various protein quantitation methods[a]

Protein	Folin Phenol	Coomassie Brilliant Blue G-250	UV Absorption (nm)			OPA Fluorometric	
			280	224–236	205	Intact	Hydrolyzed
1 Bovine serum albumin	1.00	1.00	1.00	1.00	1.00	1.00	1.00
2 Chymotrypsinogen A	1.52	0.58	2.89	1.51	1.14	0.65	1.09
3 Trypsin	1.34	0.34	1.89	1.34	1.07	0.57	1.02
4 Soybean trypsin inhibitor	0.83	0.66	1.04	0.93	1.00	0.65	0.98
5 γ-Globulin	1.07	0.46	1.51	0.98	1.04	0.77	...
6 Cytochrome *c*	1.39	1.20	2.28	1.07	1.14	0.85	0.98
7 Ovalbumin	0.93	0.52	0.88	0.83	0.98	0.64	0.74
8 Myoglobin	0.84	1.38	3.07	1.47	1.16	2.14	1.21
9 Ribonuclease A	1.28	0.68	0.84	0.76	0.99	0.77	0.86
10 Lysozyme	1.54	1.00	4.19	2.09	1.23	0.70	0.88
11 Protamine sulfate	0.90	2.15	0.02	0.41	0.91	0.03	0.89
12 Calf thymus histones	1.24	1.10	0.96	0.72	1.00	1.12	0.96
13 Calf thymus HMG proteins	1.51	1.16	2.45	0.87	1.11	1.61	1.03
14 Mixture of Nos. 1–10	1.36	0.85	3.26	1.22	1.17	0.77	1.00
15 Pig heart membranes	0.81	0.54	2.30	0.94	1.10	0.83	0.86
$\overline{X}$ (Nos. 1–10)	1.17	0.78	2.01	1.20	1.07	0.87	0.97
± SD	±0.28	±0.34	±1.16	±0.41	±0.09	±0.46	±0.14
$\overline{X}$ (Nos. 1–13)	1.18	0.94	1.81	1.08	1.06	0.88	0.97
± SD	±0.27	±0.49	±1.19	±0.43	±0.09	±0.52	±0.12

[a] The response relative to that of bovine serum albumin was determined in triplicate for 25 μg protein in the standard assay for the Folin phenol and Coomassie Brilliant Blue G-250 assay procedures, at 4–5 concentrations between 25 μg/mL and 1 mg/mL for the UV absorption methods, and in triplicate at 1000 and 50 ng for the intact and hydrolyzed protein *o*-phthalaldehyde (OPA) fluorometric assays, respectively. Hydrolysis was carried out in sealed tubes for 16 hr at 120°C.

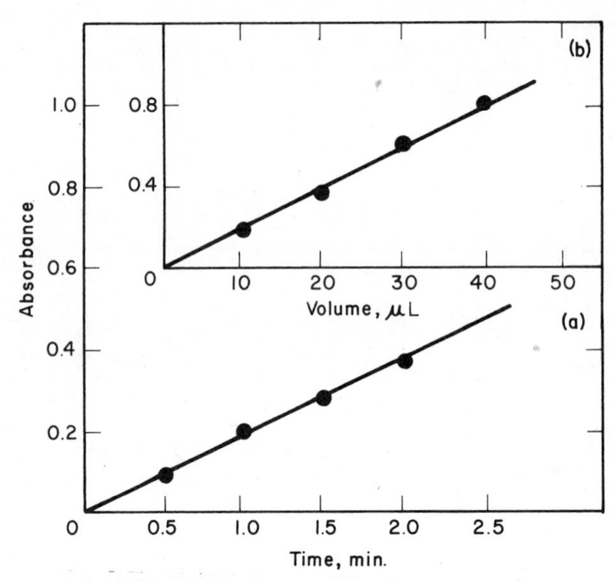

Fig. 2.2. Criteria for a valid enzyme assay. (*a*) shows a linear increase in absorbance with time for an enzyme catalyzed reaction. (*b*) shows the desired linear response of the rate of the enzyme catalyzed reaction as a function of the volume in microliters (μL) of the enzyme sample.

catalyzes a reaction under defined conditions. The rate of the reaction is obtained from the initial slope of a graph or recorder tracing showing the concentration of product formed or substrate remaining as a function of time. To be valid, an enzyme assay must be linear over the time period of the measurement [Figure 2.2(*a*)] and show a proportional response of the rate to increasing enzyme concentration [Figure 2.2(*b*)].

It is critical that the reaction conditions be regulated so that assays can be compared from one day to the next or from one laboratory to another. First, the temperature must be regulated. Make sure the substrate is at the correct temperature before initiating the reaction by adding enzyme. Adding 50 μL of a cold enzyme solution to a 1 mL or larger reaction mixture will not change the temperature sufficiently to affect the initial rate.

Other conditions that must be controlled other than substrate concentration, which will be discussed later, are the pH and ionic strength. Buffer concentration should be high enough to maintain the pH particularly if protons are taken up or produced during the reaction. The ionic strength must be consistent with optimal catalytic activity.

Units of Enzyme Activity

There are two types of units commonly used to express the amount of enzyme in a sample. The International Unit (IU), the most common, was established in 1961 by the Enzyme Commission of the International Union of Biochemistry and is defined as the amount of enzyme that transforms 1 μmol substrate/min under defined reaction conditions. Less often used is a unit known as the katal, established in 1973 by the Commission on Biological Nomenclature; it is defined as the amount of enzyme that transforms 1 mmol of substrate in 1 sec. The IU will be used in this text.

The specific activity of an enzyme is the total units of enzyme activity in a sample divided by the total protein in the sample.

$$\text{specific activity} = \text{units/mg or} \quad \mu\text{mol/min mg}$$

It depends on the reactions' conditions such as the temperature, pH, ionic strength, buffer, and concentration of substrates and cofactors.

Recall from the reaction rate law that an increase in temperature increases the rate of a reaction. A higher temperature, therefore, increases the sensitivity of the enzyme reaction. However, as the temperature is raised, the rate of denaturation of an enzyme also increases. For some enzymes, this denaturation rate is already significant at 37°C. I recommend 30°C as an assay temperature because most enzymes are stable for extended periods of time at this temperature, and water baths are easier to regulate at 30°C than at lower temperatures.

The remaining parameters of an enzyme reaction—pH, buffer, ionic strength, and concentrations of substrates and cofactors—are often interrelated. For instance, changing the pH of a reaction may increase the K_m for the enzyme, therefore requiring a higher concentration of substrate for maximum velocities. If the pH optimum of the reaction and the K_m values for the substrates are not available from the literature, select a pH near 7 and proceed by trial and error to estimate the concentration of substrate and cofactors giving an optimal rate of reaction. Using this substrate concentration, estimate an apparent pH optimum by determining the rate of the reaction in various buffers spanning the range from pH 3 to 9 using increments of 0.5 pH units. Procedures for determining a better estimate of the K_m value are discussed in Chapter 5.

If the K_m value(s) for an enzyme reaction is (are) known, select a substrate concentration equivalent to at least 10 times the K_m value(s). By substituting $S = 10\ K_m$ into the Michaelis–Menten equation,

$$v = V[S]/(K_m + [S]) = 0.91\ \ V$$

the calculated initial velocity v will then be near 91% of the maximum velocity V. If the substrate is expensive, lower concentrations can be used but at the expense of sensitivity and accuracy. One precaution should be noted. Some enzyme reactions are inhibited by their substrate at high concentrations (10-K_m levels) in which case reduced substrate concentrations may give higher reaction rates and thus a more sensitive enzyme assay.

ENZYME ASSAY

Direct Enzyme Assay

Two basic types of direct enzyme assays are in general use: end point assays and continuous assays. For an end point assay, an enzyme reaction is allowed to proceed for a fixed period of time before being stopped by rapid addition of a protein denaturant. The amount of product is then determined and expressed as micromoles per minute per unit volume of enzyme solution or unit weight of protein. By carefully planning the experimental design, end point assays can be used to determine enzyme concentration in a large number of samples such as column fractions in a convenient and rapid way. Determining the product concentration in the stopped reaction mixture, improves the economy of the analysis.

In the continuous enzyme assay, the decrease in the concentration of a substrate or an increase in product concentration is observed continuously over time. Changes in the concentration of substrate or product are usually measured by absorption or fluorescence spectroscopy, but other sensing devices, such as a pH Controller (stat), specific electrodes, and manometers may be employed as well.

Coupled Enzyme Assay

Many enzyme reaction products cannot be determined directly by available technology, or their concentrations may be difficult to assess in an end point assay. Such reactions are often more conveniently followed with a coupled enzyme assay where one of the products of the enzyme to be assayed is acted upon by another enzyme (the coupling enzyme) in the reaction mixture.

For example, hexokinase may be continuously assayed by at least two different coupled enzyme systems (43), one involving one coupling enzyme, and the other, two coupling enzymes. The one coupling enzyme system uses glucose-6-*P* dehydrogenase as the coupling enzyme

$$\text{MgATP} + \text{glucose} \xrightleftharpoons{\text{hexokinase}} \text{MgADP} + \text{glucose-6-}P$$

where MgATP is the Mg^{2+} complex of adenosine triphosphate and MgADP, the Mg^{2+} complex of adenosine diphosphate. This reaction is coupled to

$$\text{glucose-6-}P + \text{NADP}^+ \xrightleftharpoons[]{\text{glucose-6-}P\text{ dehydrogenase}} \text{NADPH} + \text{6-}P\text{-gluconolactone} + \text{H}^+$$

where $NADP^+$ and NADPH are the oxidized and reduced forms, respectively, of nicotinamide adenine dinucleotide phosphate. In the second system, the hexokinase reaction is coupled to both pyruvate kinase and lactate dehydrogenase:

$$\text{MgATP} + \text{glucose} \xrightleftharpoons{\text{hexokinase}} \text{MgADP} + \text{glucose-6-}P$$

is coupled to

$$\text{MgADP} + P\text{-enolpyruvate} \xrightleftharpoons[\text{K}^+]{\text{pyruvate kinase}} \text{MgATP} + \text{pyruvate}$$

and

$$\text{H}^+ + \text{pyruvate} + \text{NADH} \xrightleftharpoons{\text{lactate dehydrogenase}} \text{NAD}^+ + \text{lactate}$$

where NAD^+ and NADH are the oxidized and reduced forms, respectively, of nicotinamide adenine dinucleotide. Because NAD^+ or $NADP^+$ dependent dehydrogenases are conveniently monitored, most investigators prefer to use them in coupled enzyme reactions. It is, therefore, important to appreciate the properties of these pyridine nucleotide coenzymes given in Inset 2.2.

Occasionally, enzymes are assayed by coupling to nonenzymic reactions (46), but in all cases, that is, in both the enzymic and nonenzymic coupled systems, it is important to ensure that the observed rate of the reaction is not limited by the coupling system. The rate of the coupled reaction must be directly proportional to the concentration of the enzyme being assayed.

Since coupling enzymes are expensive, it may be important to minimize the cost by adding the minimum amount of the coupling enzyme(s) to each reaction mixture. Of interest then is the question: How does one determine the minimum amount of each coupling enzyme to add without destroying the validity of the assay?

Reaction (2.4),

$$A \xrightarrow[\text{enzyme 1 (primary)}]{v_1} B \xrightarrow[\text{C} \nearrow \text{ enzyme 2 (coupling)} \searrow]{v_2} P + Q \quad (2.4)$$

represents a typical coupled enzyme assay. The terms v_1 and v_2 are the rates of the primary and coupling enzymes, respectively. Because NAD^+ and $NADP^+$ dependent dehydrogenases are often used as coupling enzymes, enzyme 2 is represented by

$$B + C \rightleftharpoons P + Q.$$

where C and Q represent oxidized and reduced pyridine nucleotide coenzymes. Assume for the following treatment that the concentration of A is much larger than K_{mA}, its Michaelis constant, and that the initial velocity of enzyme 1, v_1, is constant during the assay. The rate of the coupling enzyme, v_2, is given by

$$v_2 = V_2[B]/(K_{mB} + [B]) \quad (2.5)$$

where V_2 is the apparent maximal velocity of the coupling enzyme at finite concentrations of the coenzyme C and K_{mB} is the Michaelis constant for B. The relationship between V_2 and the velocity of the coupling enzyme at infinite concentrations of B and C, V_{2max}, is given by

$$V_2 = V_{2(max)}[C]/(K_{mC} + [C]) \quad (2.6)$$

INSET 2.2

Some Useful Properties of the Pyridine Nucleotide Coenzymes NAD^+, NADH, $NADP^+$, and NADPH

The following data are taken from Reference (44).

	pH	Absorption Maximum (nm)	ϵ (X10^{-3}) (M^{-1} cm^{-1})	Absorption Minimum (nm)	ϵ (X10^{-3}) (M^{-1} cm^{-1})	Fluorescence Emission (nm)	Molecular Weight
NAD^+	7.5	259	17.8	230	8.0		664
$NADP^+$	7.0	259	18.0	231	8.1		743
NADH	7.5	259	14.4	234	6.6	457	665
		338	6.22	290	1.3		
NADPH	7.5	259	16.9	236	7.6	457	744
		339	6.2	290	1.4		

The oxidized pyridine nucleotides NAD^+ and $NADP^+$ have essentially the same extinction coefficients; likewise, the reduced forms also have the same extinction values. Both NADH and NADPH fluoresce with emission maxima at 457 nm when excited at 340 nm. The oxidized coenzymes do not fluoresce.

The most convenient method for determining the concentration of pyridine nucleotide coenzymes is to measure their absorbances. However, because many pyridine nucleotide preparations contain contaminating hydrolytic products that absorb light at the absorbance maxima of the desired cofactor and/or they contain significant quantities of the inactive α isomer, the spectrophotometric method is not the most accurate. A more accurate procedure for determining the concentration of these cofactors utilizes enzymic procedures. The enzymic method takes advantage of the changes in absorbance at 340 nm when an enzyme reaction is initiated by adding an aliquot of the oxidized or reduced coenzyme. An accurate determination, however, requires that the proper pH and substrate concentrations be chosen in order for the reaction to go to completion.

The stability of the pyridine nucleotide coenzymes is one of their least appreciated properties. In general, the oxidized coenzymes are stable under acidic conditions but not basic conditions. The reduced forms have opposite properties: They are stable under basic conditions but not acidic conditions. Under basic conditions (pH 10) many nucleophilic reagents such as cyanide form reversible adducts with the oxidized forms by reacting at the para position of the nicotinamide moiety. High concentrations of the hydroxyl ion also add to the 2 or 4 position of the nicotinamide ring to cause ring opening and other hydrolytic events. The reduced coenzymes are destroyed rapidly by acid; $t_{1/2}$ at pH 3 and 25°C is 15 min (45). More importantly, general acids such as inorganic phosphate at pH 7.0, catalyze an irreversible loss of the coenzyme function. As a general precaution then, store NADH and NADPH solutions in amine buffers at pH 7.5–8.5.

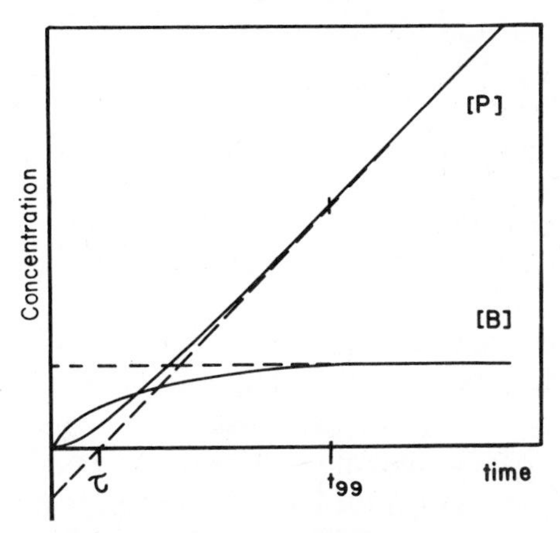

Fig. 2.3. Relationship between product concentration and time for a hypothetical enzyme assay. The solid line designated with [P] represents the time course for the appearance of product: The dashed line on the [P] curve is an extrapolation from the steady state rate for the appearance of product to the time axis. The intersection of the extrapolated dashed line with the time axis defines Easterby's τ value (48). The solid line designated [B] represents the time course for the appearance of the intermediate [B]: the dashed line from the [B] curve represents the concentration of B in the steady state. The time in which the concentration of the intermediate is 99% of its steady state value is t_{99}. Reproduced by permission of the National Research Council of Canada from Brooks et al., *Can. J. Biochem. Cell Biol.*, **62**, 945–955 (1984).

where K_{mC} is the Michaelis constant for C (47). For this analysis, V_2 is expressed in concentration terms, units/assay or units/mL; its value may be changed either by adding more coupling enzyme to the assay or by increasing the concentration of C as shown by Equation 2.6.

A close examination of reaction (2.4) reveals that the concentration of intermediate B must increase from zero to some finite value after the reaction is initiated. After the velocity of the coupling enzyme equals the velocity of the primary enzyme as portrayed in Figure 2.3, the concentration of B no longer changes with time; it reaches its steady state value $[B]_{ss}$. Note that the rate of formation of product P is not linear from time zero; its rate of formation also increases with time until the concentration of B reaches its steady state value. The t_{99} value, indicated on Figure 2.3, is the lag time required for B to reach 99% of its steady state concentration, which is equivalent to the time it takes for the rate of formation of product P to reach 99% of its final linear rate.

The rate of change of the concentration of B, after a coupled enzyme reaction is initiated, is the difference between its rate of formation and its rate of breakdown as defined by Equation (2.7) (49).

$$d[\mathrm{B}]/dt = v_1 - V_2[\mathrm{B}]/(K_{m\mathrm{B}} + [\mathrm{B}]) \tag{2.7}$$

Integration of Equation (2.7), holding v_1 and V_2 constant with respect to time, and assuming a boundary condition of $[\mathrm{B}] = 0$ at $t = 0$, gives

$$\{K_{m\mathrm{B}}v_1 - [\mathrm{B}](V_2 - v_1)\} \exp \{[\mathrm{B}](V_2 - v_1)/K_{m\mathrm{B}}V_2\} = K_{m\mathrm{B}}v_1 \exp [-\mathrm{t}(V_2 - v_1)^2/K_{m\mathrm{B}}V_2] \tag{2.8}$$

Equation (2.8), the general equation for a one coupled enzyme system, describes the concentration of B at any time t after initiation of the reaction. In order to calculate the elapsed time before the observed velocity is approximately equal to the rate of the primary enzyme, a value for F_B must be selected, where F_B is the concentration of B divided by its steady state concentration $[\mathrm{B}]_{ss}$:

$$[\mathrm{B}]_{ss} = v_1 K_{m\mathrm{B}}/(V_2 - v_1) \tag{2.9}$$

Because $[\mathrm{B}]_{ss}$ is the concentration of B when $d[\mathrm{B}]/dt = 0$ [see Equation (2.7)], it follows that

$$F_\mathrm{B} = [\mathrm{B}]/[\mathrm{B}]_{ss} = [\mathrm{B}](V_2 - v_1)/v_1 K_{m\mathrm{B}} \tag{2.10}$$

Substituting Equation (2.10) into Equation (2.8) gives an expression for t_{F_B} for a coupled enzyme reaction containing v_1 and V_2 units of the primary and coupling enzyme, respectively,

$$t_{F_\mathrm{B}} = [-K_{m\mathrm{B}}/(V_2 - v_1)^2][F_\mathrm{B}v_1 + V_2 \ln (1 - F_\mathrm{B})] \tag{2.11}$$

When the concentration of coupling enzyme is such that $(K_{m\mathrm{B}} + [\mathrm{B}]) \simeq K_{m\mathrm{B}}$, Equation (2.11) reduces to

$$t_{F_\mathrm{B}} = (-K_{m\mathrm{B}}/V_2) \ln (1 - F_\mathrm{B}) \tag{2.12}$$

Rearranging Equation (2.11) gives Equation (2.13), which enables one to calculate V_2 for a selected lag time

$$V_2^2 + V_2[K_{m\mathrm{B}} \ln (1 - F_\mathrm{B})/t_{F_\mathrm{B}} - 2v_1] + v_1^2 + F_\mathrm{B}K_{m\mathrm{B}}v_1/t_{F_\mathrm{B}} = 0 \tag{2.13}$$

When calculating V_2, be sure $K_{m\mathrm{B}}$ and v_1 are given in the same concentration units. Figure 2.4 shows the lag times calculated with Equations (2.11) and (2.12) for various ratios of v_1/V_2. Note that when v_1/V_2 is less

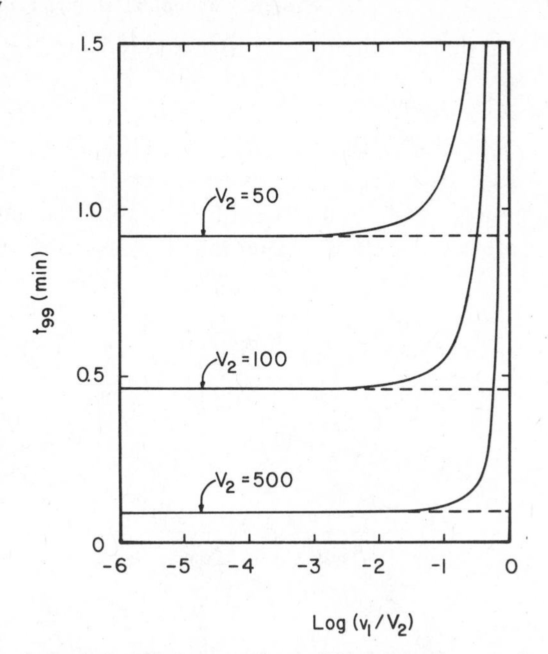

Fig. 2.4. Theoretical relationship between log (v_1/V_2) and the time required for [B] to reach 99% its steady state value calculated with Equation (2.11) (solid curve) and Equation (2.12) (dashed curve) assuming $K_{mB} = 10\ \mu M$ and a 1-mL reaction volume. Each curve represents a different value of V_2 as shown on the figure. Reproduced by permission of the National Research Council of Canada from Brooks et al., *Can. J. Biochem. Cell Biol.*, **62**, 945–955 (1984).

than 0.01, that is, when $\log(v_1/V_2)$ is greater than -2, there is considerable error in the t_{99} lag time calculated with Equation (2.13) (solid line). At values of log (v_1/V_2) less than -2, both Equations (2.11) and (2.12) give the same result.

It is important to know that the longer the lag time, the more substrate is consumed before the steady state is achieved. This precaution is particularly important if the substrate concentration of enzyme 1 is below its K_m value. This is required when K_m values are to be determined. High concentrations of the coupling enzymes are then desirable, $V_2 = 100\ v_1$.

For a system requiring two coupling enzymes,

$$\mathrm{A} \xrightarrow[\text{primary enzyme}]{v_1} \mathrm{B} \xrightarrow[\text{coupling enzyme 1}]{v_2} \underset{\mathrm{D}}{\mathrm{C}} \xrightarrow[\text{coupling enzyme 2}]{v_3} \underset{\mathrm{Q}}{\mathrm{P}}$$

it follows that

$$d[\mathrm{C}]/dt = V_2[\mathrm{B}]/(K_{mB} + [\mathrm{B}]) - V_3[\mathrm{C}]/(K_{mC} + [\mathrm{C}]) \qquad (2.14)$$

When $d[\mathrm{C}]/dt = 0$, that is, when $[\mathrm{B}] = [\mathrm{B}]_{ss}$ and $[\mathrm{C}] = [\mathrm{C}]_{ss}$, the velocity is given by

$$v_{\mathrm{obs}} = V_3[\mathrm{C}]_{ss}/(K_{\mathrm{mC}} + [\mathrm{C}]_{ss})$$

Because Equation (2.14) cannot be integrated by conventional methods, Brooks et al. (47) developed Equations (2.15) and (2.16) to approximate $t_{F\mathrm{C}}$.

$$t_{F\mathrm{C}} = -\frac{K_{\mathrm{mB}}}{(V_2 - v_1)} \ln\left\{1 - \frac{K_{\mathrm{mC}}(V_2 - v_1)}{K_{\mathrm{mB}}(V_3 - v_1)}(1 - F_{\mathrm{C}})\right\} \quad (2.15)$$

when

$$\frac{2(V_2 - v_1)}{K_{\mathrm{mB}}} \leq \frac{(V_3 - v_1)}{K_{\mathrm{mC}}}$$

or

$$t_{F\mathrm{C}} = -\frac{K_{\mathrm{mC}}}{(V_3 - v_1)} \ln\left\{1 - \frac{K_{\mathrm{mB}}(V_3 - v_1)}{K_{\mathrm{mC}}(V_2 - v_1)}(1 - F_{\mathrm{C}})\right\} \quad (2.16)$$

when

$$\frac{(V_2 - v_1)}{K_{\mathrm{mB}}} \geq \frac{2(V_3 - v_1)}{K_{\mathrm{mC}}}$$

If

$$(V_2 - v_1)/K_{\mathrm{mB}} = (V_3 - v_1)/K_{\mathrm{mC}}$$

then Equations (2.15) and (2.16) fail, so that the approximate values of $t_{F\mathrm{C}}$ can only be calculated after decreasing the value for either V_2 or V_3 by 1%. A better estimate of $t_{F\mathrm{C}}$ than that given by Equations (2.15) and (2.16) is obtained with Equation (22) of Reference (47), although computer assistance is necessary. The estimates given by Equations (2.15) and (2.16) are always less but usually by no more than 5% of the true value.

The difficulty of using Equation (2.15) or Equation (2.16) is that their solution requires a prior knowledge of V_2 and V_3. Furthermore, several combinations of V_2 and V_3 values will give the same $t_{F\mathrm{C}}$ value. For the two enzyme case then, the usual procedure is to select a desired $t_{F\mathrm{C}}$ and

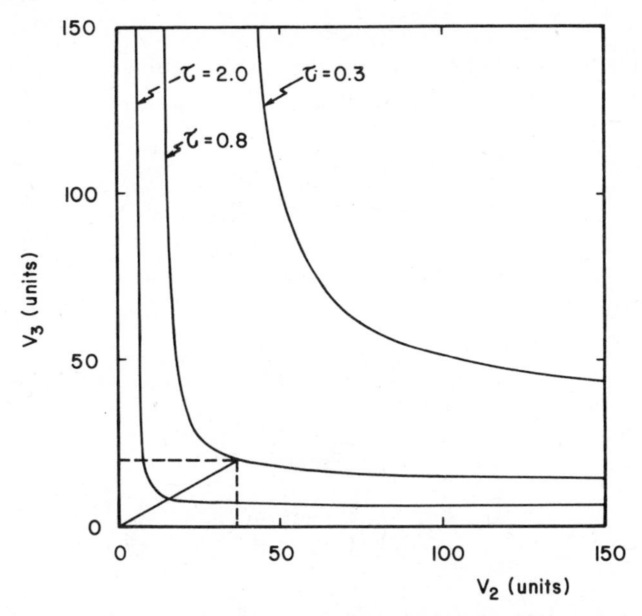

Fig. 2.5. Graphical representation of the cost minimization technique. The three curves, generated by Equation (2.18), represent the relationship between V_2 and V_3 for various values of τ. The slope of the straight solid line is defined by κ from Equation (2.17) with $P_2/P_3 = 3$ and $\tau = 0.8$ min. The dotted lines are extrapolations of the point of intersection of the functions from Equations (2.18) and (2.17) to the axis. A reaction volume of 1 mL is assumed. Reproduced by permission of the National Research Council of Canada from Brooks et al., *Can. J. Biochem. Cell Biol.*, **62**, 945–955 (1984).

then determine the optimum amount of V_2 and V_3 required to achieve the defined lag time.

To define minimum values of V_2 and V_3, Cleland (43) chose a function that minimizes the total concentration of the two coupling enzymes with respect to the price per unit of each enzyme. However, the functions that describe the lag time are not explicit for V_2 and V_3, and, therefore it is necessary to first minimize the cost for a selected transition time τ (see Figure 2.3). This minimization procedure leads to

$$\kappa = \frac{V_2}{V_3} = \frac{P_3P_2(\tau v_1 + K_{mB}) + P_3(P_2P_3K_{mC}K_{mB})^{1/2}}{P_3P_2(\tau v_1 + K_{mC}) + P_2(P_2P_3K_{mC}K_{mB})^{1/2}} \tag{2.17}$$

where κ is the slope of the line of minimum cost as depicted in Figure 2.5 and P_2 and P_3 are the price per unit of enzyme 2 and 3. The minimum values of V_2 and V_3 required to give a specified τ value are defined by the point of the intersection of the line with slope κ and the curves in Figure 2.5. These curves, calculated with Equation (2.18), give V_3 as a function of V_2 at constant τ values

$$V_3 = \frac{K_{mC}(V_2 - v_1)}{(\tau(V_2 - v_1) - K_{mB})} + v_1 \qquad (2.18)$$

The point at which the straight line and the curve intersect on Figure 2.5 is given by

$$V_2 = (R_1 + (R_2^2 + \kappa K_{mB} K_{mC})^{1/2})/2\tau \qquad (2.19)$$

where

$$R_1 = v_1\tau + K_{mB} + \kappa(K_{mC} + \tau v_1)$$
$$R_2 = v_1\tau + K_{mB} - \kappa(K_{mC} + \tau v_1)$$

and V_3, defined by Equation (2.17), equals $V_2\, \kappa$.

Although this procedure minimizes V_2 and V_3, the fixed τ value has no practical use. However, an approximate linear relationship exists between t_F and τ (46) as defined by

$$t_{99} \simeq 4.28\tau \qquad (2.20)$$

and

$$t_{90} = 2.79\tau \qquad (2.21)$$

Because the actual values of t_F vary considerably from the above relationships, Equations (2.20) and (2.21) serve only as an approximate guide. Once V_2 and V_3 values are calculated, better estimates of the lag times are obtained by substituting these values into Equation (2.15) or (2.16) or with Equation (22) from Brooks et al. (47).

This theoretical analysis of coupled enzyme reactions does not consider the effect of mutarotation of products on the coupling parameters. Glucose 6-phosphate, an intermediate product in both of the coupling systems described earlier in this chapter, undergoes mutarotation at the carbon 1 position giving a mixture of the α and β enantiomers. Because only the β enantiomer is a substrate for glucose 6-phosphate dehydrogenase, the lag time will depend not only on the kinetic parameters of the coupling enzymes but also on the rate of interconversion of the α and β enantiomers (43). It is therefore essential that the stereochemical configuration of the intermediates in a coupled enzyme reaction be carefully examined to ensure that they do not affect the coupling parameters. The equations that predict the lag times for these reactions are given in Reference 50. A computer program written for an IBM PC with 64K RAM, called Lagtime, is available for a modest charge from the Instructional Media Center, Marketing Division, Michigan State University, East Lansing MI, 48824. The program can be used to calculate the units of enzyme required to establish a valid coupled assay with a defined lagtime or the lagtime expected when the units of coupling enzyme are known.

Initiating a Reaction

In general, a small volume (10–25 μL) of a concentrated enzyme solution should be added to the complete reaction mixture to initiate the enzyme reaction. This procedure has at least two advantages: first, the enzyme will not remain diluted at the assay temperature for extended periods of time prior to the start of the reaction and second, small volumes can be added in less time than larger volumes so that the initial velocities can be measured as soon as possible after addition of the enzymes.

Difficulties with Enzyme Assays

In addition to the usual difficulties of measuring rates of chemical reactions, enzyme reaction rate assays have other problems associated with them. Crude extracts often contain contaminating enzyme activities that utilize either a product or a reactant of the enzyme of interest. Sometimes these contaminating enzymes can be inhibited; NADH oxidase, for instance, can be inhibited with cyanide. In other cases, unknown activators or inhibitors in the crude extract influence the rates. Consequently, correct estimates of the units of enzyme activity in an extract may not be possible to obtain until after the first or second step of the purification scheme.

The loss of enzyme activity during the assay is another possible problem, although it is more common when near homogeneous enzyme rather than crude homogenates are assayed. Because small amounts of protein are required for an assay when the enzyme is homogeneous, a nonlinear response of enzyme activity with time and/or concentration may be due to its inactivation either by dilution or by adsorption to the reaction vessel. Bovine serum albumin (1 mg/mL), 20% glycerol (v/v), or 0.2 m*M* Triton X-100 may be included in the assay or the reaction conditions may be modified in other ways in an attempt to stabilize the enzyme; using higher concentrations of the enzyme may also help. A more extended discussion of enzyme inactivation or denaturation is provided in Chapter 1.

REFERENCES

1. G. L. Peterson, *Methods Enzymol.,* **91,** 95–119 (1983).
2. C. J. R. Thorne, Techniques for determining protein concentration, in H. L. Kornberg, J. C. Metcalfe, D. H. Northcote, C. I. Pogson, and K. F. Tipton, Eds, *Techniques in Protein and Enzyme Biochemistry,* Pt. 1, B104, Elsevier-North Holland, Amsterdam, 1978, pp. 1–18.
3. R. K. Scopes, *Anal. Biochem.,* **59,** 277–282 (1974).
4. E. Layne, *Methods Enzymol.,* **3,** 447–454 (1957).
5. J. R. Whitaker and P. E. Granum, *Anal. Biochem.,* **109,** 156–159 (1980).
6. W. E. Groves, F. C. Davis, Jr., and B. H. Sells, *Anal. Biochem.,* **22,** 195–210 (1968).
7. P. Wolf, *Anal. Biochem.,* **129,** 145–155 (1983).
8. M. A. K. Markwell, S. M. Haas, N. E. Tolbert, and L. L. Bieber, *Methods Enzymol.,* **72,** 296–303 (1981).
9. J. V. Castell, M. Cervera, and R. Marco, *Anal. Biochem.,* **99,** 379–391 (1979).
10. S. M. Read and D. H. Northcote, *Anal. Biochem.,* **116,** 53–64 (1981).
11. O. Warburg and W. Christian, *Biochem. Z.,* **310,** 384–421 (1941).
12. W. E. Cohn, Molar absorptivity and $A_{1cm}^{1\%}$ values for proteins at selected wavelengths of the ultraviolet and visible region, in G. D. Fasman, Ed., *Handbook of Biochemistry and Molecular Biology, Proteins,* Vol. II, 3rd Ed., CRC Press, Cleveland, 1975, pp. 383–545.
13. P. Welch and R. K. Scopes, *Anal. Biochem.,* **112,** 154–157 (1981).
14. O. H. Lowry, N. J. Rosebrough, A. L. Farr, and R. J. Randall, *J. Biol. Chem.,* **193,** 265–275 (1951).
15. Science Citation Index, Jan.–Dec. 1981. Institute for Scientific Information, Philadelphia PA.
16. G. L. Peterson, *Anal. Biochem.,* **100,** 201–220 (1979).
17. H.-P. Schroer and K. Augsten, *Z. Chem.* **18,** 391–392 (1978).
18. V. Chromy, J. Fischer, and V. Kulhanek, *Clin. Chem.,* **20,** 1362–1363 (1974).
19. A. Bensadoun and D. Weinstein, *Anal. Biochem.,* **70,** 241–250 (1976).
20. G. L. Peterson, *Anal. Biochem.,* **83,** 346–356 (1977).
21. I. Polacheck and E. Cabib, *Anal. Biochem.,* **117,** 311–314 (1981).
22. P. J. Geiger and S. P. Bessman, *Anal. Biochem.,* **49,** 467–473 (1972).
23. A. Rengasamy and A. Gnanam, *Ind. J. Biochem. Biophys.,* **18,** 67–68 (1981).

24. J. Hughes, S. Joshi, and D. Ascoli, *Anal. Biochem.*, **117**, 1–5 (1981).
25. H. P. S. Makkar, O. P. Sharma, and S. S. Negi, *Anal. Biochem.*, **104**, 124–126 (1980).
26. W. T. Coakley and C. J. James, *Anal. Biochem.*, **85**, 90–97 (1978).
27. V. Chromy, J. Fischer, and J. Voznicek, *Z. Med. Labor.-Diagn*, **21**, 333–337 (1980).
28. W.-T. Law and S. R. Crouch, *Anal. Lett.*, **13**, 1115–1128 (1980).
29. S. T. Ohnishi and J. K. Barr, *Anal. Biochem.*, **86**, 193–200 (1978).
30. M. M. Bradford, *Anal. Biochem.*, **72**, 248–254 (1976).
31. J. Pierce and C. H. Suelter, *Anal. Biochem.*, **81**, 478–480 (1977).
32. H. van Kley and S. M. Hale, *Anal. Biochem.*, **81**, 485–487 (1977).
33. Z. Zaman and R. L. Verwilghen, *Anal. Biochem.*, **109**, 454–459 (1980).
34. G. S. McKnight, *Anal. Biochem.*, **78**, 86–92 (1977).
35. R. Almog and D. S. Berns, *Anal. Biochem.*, **114**, 336–341 (1981).
36. H. Nakamura and J. J. Pisano, *Arch. Biochem. Biophys.*, **172**, 102–105 (1976).
37. R. Doolittle, *Science*, **214**, 149–159 (1981).
38. P. E. Hare, *Methods Enzymol.*, **47**, 3–18 (1977).
39. K. Nishio and M. Kawakami, *Anal. Biochem.*, **126**, 239–241 (1982).
40. W. Mejbaum-Katzenellenbogen and W. M. Dobryszycka, *Clin. Chim. Acta*, **4**, 515 (1959).
41. S. B. Pruett and M. Wolcott, *J. Immunol. Methods*, **35**, 129–136 (1980).
42. A. K. Hazra, S. P. Chock, and R. W. Albers, *Anal. Biochem.*, **137**, 437–443 (1984).
43. W. W. Cleland, *Anal. Biochem.*, **99**, 142–145 (1979).
44. J. G. Morris and E. R. Redfearn, Vitamins and Coenzymes, in R. M. C. Dawson, D. C. Elliott, W. H. Elliott, and K. M. Jones, Eds., *Data for Biochemical Research*, 2nd ed., Oxford University Press, New York, 1969, pp. 191–215.
45. C. C. Johnston, J. L. Gardner, C. H. Suelter, and D. E. Metzler, *Biochemistry*, **2**, 689–696 (1963).
46. P. A. Srere, H. Brazil, and L. Gonen, *Acta Chem. Scand.*, **17**, s129–s134 (1963).
47. S. P. J. Brooks, T. Espinola, and C. H. Suelter, *Can. J. Biochem. Cell Biol.*, **62**, 945–955 (1984).
48. J. S. Easterby, *Biochim. Biophys. Acta*, **293**, 552–558 (1973).
49. W. R. McClure, *Biochemistry*, **8**, 2782–2786 (1969).
50. S. P. J. Brooks, T. Espinola, and C. H. Suelter, *Can. J. Biochem. Cell Biol.*, **62**, 956–963 (1984).

3

Purification of an Enzyme

Early in the development of enzymology, biochemists learned that to understand the structure and function of enzymes it was necessary to first purify them. Purification of enzymes was necessary to prove that they were proteins. While no one today doubts that they are, for numerous research, medical, and industrial reasons, it is still necessary to purify and characterize them. This chapter, therefore, presents a review of the arsenal of tools available to enzymologists to purify proteins. Subsequent chapters will deal with their characterization.

STRATEGY FOR DEVELOPING A PURIFICATION SCHEME

Enzymologists today have a much better understanding of the effects of various salts, solvents, and environmental conditions on polar and nonpolar forces that dictate the behavior of proteins in solution than they did 25 years ago. In recent years, therefore, they have developed a large number of methods to fractionate proteins. These separation methods are based on differences in (a) molecular size, (b) electrostatic and nonpolar properties, and (c) substrate and inhibitor specificities of each protein. As a result of the development of these methods, the purification of an enzyme is now a science and not an art as it once was. This is not to imply that a unique purification scheme exists for each protein; no doubt several schemes can be developed.

A flow diagram for a purification scheme is outlined in Figure 3.1. Before beginning the development of such a scheme, one needs a suitable assay for the enzyme (see Chapter 2), and a tissue, which contains the

Prepare a crude extract
Fractionate crude extract
Characterize enzyme
Purify enzyme by
- Denaturing contaminating protein
- Gel permeation chromatography
- Ion exchange chromatography
- Hydrophobic chromatography
- Dye ligand affinity chromatography
- Bio-ligand affinity chromatography
- High performance liquid chromatography
- Fast protein liquid chromatography
- Electrophoretic separation methods
    - Electrophoresis
    - Electrofocusing
- Miscellaneous fractionation methods
- Crystallization

Fig. 3.1. A typical flow diagram for the purification of an enzyme.

enzyme. Time spent developing an efficient assay before proceeding with the development of the purification scheme will pay dividends later because column and other fractions must be assayed for enzyme activity at every step. After an assay is developed, a tissue selected, and a crude extract prepared, some of the properties of the enzyme in the crude extract are measured. These data are used to select the order in which the different fractionation methods should be applied.

PREPARING A CRUDE EXTRACT

Choosing a Tissue

Choosing a tissue may or may not be critical for the development of a purification scheme. If an enzyme from a specific tissue is required, there is no choice to make. In other cases, a choice can be made, which is usually based on economic criteria, that is, the enzyme is isolated from the tissue containing the highest concentration. Sometimes, however, the amount of enzyme in a tissue can be enriched by metabolite induction, by mutation to constitutive levels (1), or by genetic manipulation using recombinant deoxyribonucleic acid (DNA) techniques (2). Because many enzymes are compartmentalized in organelles and membranes, the isolation of the organelle, as a first step, would significantly enrich the enzyme.

Whatever the tissue source, there are often difficulties associated with the isolation of an enzyme from it. Unless the enzyme is extracellular, like those in serum, in an intestinal tract, or in a bacterial culture medium, the tissue must first be ruptured to release it. Some tissues, however, such as yeast cells, placenta, skin, and others are difficult to disrupt. After lysis of the tissue cells is complete, other problems are often encountered. The high ratio of nucleic acid to protein in microbial tissue complicates fractionation; liver and yeast extracts contain elevated amounts of proteolytic enzymes; some plant tissues contain large quantities of phenolic compounds that interact strongly with proteins; and intrinsic membrane proteins require specific solvents to solubilize them. Each of these difficulties will be discussed in the subsequent sections as the various tissue sources are examined.

Microbial Cells

Microbial cells are ruptured by sonication, by passage through a French press (3, 4) or Manton–Gaulin homogenizer (5), by blending with glass beads (6), or by digesting the cell walls enzymically (7). Cell rupture is easily monitored by microscopic examination. The crude extract is the supernatant liquid obtained after centrifugation of the ruptured cell suspension. If the ruptured cell suspension contains a gelatinous aggregate of nucleic acids, it should be sonicated before centrifugation.

After preparing the crude extract, determine the ratio of nucleic acid to protein by measuring absorbances at 260 and 280 nm. Based on the experiments of Warburg and Christian (8), yeast nucleic acids have an $A_{280\ nm}/A_{260\ nm}$ ratio near 0.5 while yeast enolase has an $A_{280\ nm}/A_{260\ nm}$ ratio = 1.75. Hence, values of this ratio near 1 indicate a significant contamination of nucleic acid in the extract.

Nucleic acids should be removed before beginning the purification because the complexes that they form with many proteins do not fractionate cleanly. (As will be discussed later in this chapter, some crude extracts can be added directly to dye-ligand affinity matrices and perhaps to other affinity matrices without prior removal of nucleic acids.) The difficulties associated with the determination of protein in crude extracts containing a significant contamination of nucleic acids are discussed more extensively in Chapter 2.

Adding protamine sulfate (0.2–0.4% final concentration) or streptomycin (1–2%) usually precipitates nucleic acids from the crude extract. Digesting them with nucleases (6), or precipitating them with $MnCl_2$ (50 m*M*), lysozyme (12 mg/mL) (9) or 6,9-diamino-2-ethoxyacridine (Ethodin or Rivanol) (10) is also effective. An increased ionic strength [0.2 *M* $(NH_4)_2SO_4$] decreases the strength of the nucleic acid–protein interaction and may improve the efficiency of the separation. Most precipitations are done in the cold (4°C). However, it is important to know that no one reagent is universally preferable for precipitating nucleic acids from microbial extracts (9), and so it may be necessary to try several approaches before a satisfactory extract is obtained.

Yeast cells are more difficult to rupture than microbial cells; sonication and other less vigorous methods, which are effective in rupturing microbial cells do not rupture yeast cells. Inset 3.1 contains a description of three methods for rupturing yeast cells, each equally effective, but each has its own unique advantages and disadvantages. Pressure homogenization using either a Manton–Gaulin homogenizer or a French press is rapid but requires special equipment. Autolysis requires no special equipment, but it requires longer times to rupture the yeast cell than either of the other two methods. Finally, blending yeast cells frozen with liquid N_2 (14) is rapid and effective, particularly for isolating yeast mitochondria; working with liquid N_2 is its major difficulty.

The chief problem involved in using yeast as an enzyme source is the high proteinase activity in the cell extract. Yeast contains several proteinases, most of which are located in vacuoles (16); the contents of these vacuoles appear to be released as the yeast cell is ruptured. (No evidence is available comparing the amounts of proteinase released with the methods used to rupture the cells.) Unless special precautions are taken during the enzyme purification, partially degraded enzymes may be obtained.

To prevent proteolysis of an enzyme during its purification from a crude yeast extract, add a proteinase inhibitor (see Table 3.1 for list) to the yeast slurry before pressure homogenization or to the buffer that is added after autolysis or blending. Serine proteinase inhibitors in common use are phenylmethylsulfonyl fluoride, methanesulfonyl fluoride, trasylol, and diisopropylfluorophosphate. The latter inhibitor is the most effective but its toxicity often precludes its use. Ethylenediaminetetraacetic acid

INSET 3.1.

Basic Methods for Rupturing Yeast Cells

Pressure Homogenization

Homogenizing yeast by pressure is an effective way to rupture them and release the protein. Releasing protein from yeast cells is more rapid at 30°C than at 5°C and at higher pressures. The concentration of yeast, on the other hand, does not affect the rate of protein release (5, 11).

Suspend 1 lb of yeast cake in 1 L of buffer (5-m*M* Tris–HCl. 10-m*M* $MgCl_2$, and 1-m*M* dithiothreitol, pH 8.1), add a proteinase inhibitor (Table 3.1), and pass through the homogenizer (12). At 30°C and 550 kg/cm^2, 62% of the protein is released in one pass, 75% in two passes, and 95% in four passes (5). Nearly 100 mg protein is obtained from 1 g of packed yeast when the yeast is completely ruptured.

Autolysis with Toluene and 2-Mercaptoethanol

A recent modifiication (13) of an autolytic procedure using 27 mL toluene plus 0.7 mL 2-mercaptoethanol/lb of yeast is more convenient and gives better yields than the original method calling for 240 mL toluene/lb. Yeast (1 lb) is crumbled into a 2-L glass beaker and mixed with 27 mL toluene containing 0.7 mL 2-mercaptoethanol. The mixture is allowed to remain at 37°C with intermittent stirring until the yeast liquifies, usually after 90 min. Then add 265 mL of 15-m*M* EDTA containing 5 m*M* 2-mercapto-ethanol and a proteinase inhibitor and allow the solution to stir overnight at room temperature. Centrifuging removes cell wall debris from the crude yeast extract.

Blending Frozen Cells

After freezing yeast cells in liquid nitrogen (1-lb crumbled yeast in 1.5 L liquid N_2) pour liquid N_2 and frozen yeast into a stainless steel Waring blender and homogenize for 4 min at 1-min intervals. After each minute, scrape frozen yeast powder off the inner surfaces of the container. The fine frozen powder is then suspended in 20-m*M* sodium phosphate, pH 7.5, containing 2-mercaptoethanol and a proteinase inhibitor, allowed to thaw, and stirred for 1 hr (14) before centrifuging to remove cell wall debris. Cells may also be blended in dry ice (15).

Table 3.1. Useful proteinase inhibitors[a]

Inhibitor and Abbreviation	Effective Concentration (m*M*)	Comments
Serine proteinases		
Phenylmethylsulfonyl fluoride (PMSF)	1	Stable for months in 100% isopropanol. Half-life (pH 7, 25°C) = 110 min (17). Relatively nontoxic. Use with caution. Reacts with other proteins (18).
Methanesulfonyl-fluoride (MSF)	1	Also inactivates zymogens (19). Stability characteristics not available.
Dimethyl dichlorovinyl phosphate (DDVP)[b]	1	Relatively stable, half-life (pH 7.6 37°C) = 32 hr (20). Nontoxic (21) but may be a carcinogen (22).

(EDTA) and ethyleneglycoltetraacetic acid (EGTA) are useful in reducing the activity of metalloproteinases. Acid proteinase activity is reduced by maintaining the solution at pH 7 or higher. Other ways to reduce proteolysis are to use a fresh yeast cake, to reduce the time in which the crude extract is in contact with the proteinases, and to fractionate the crude extract as soon as possible after preparation. Autolytic methods requiring extensive periods of time to rupture yeast cells are usually not recommended. On the other hand, pyruvate kinase was successfully isolated from yeast after using autolysis to release it (13).

A review of the literature provides numerous examples of the isolation of partially degraded enzymes from yeast (13, 33–35). In addition, some homogeneous yeast enzyme preparations are contaminated with a proteinase. Yeast pyruvate kinase, for example, migrates as a single band on polyacrylamide gel during electrophoresis in the absence of denaturants. Yet, when this same protein is subjected to sodium dodecyl sulfate polyacrylamide gel electrophoresis, a large number of polypeptides, ranging in molecular weight from 10,000 to 50,000, are observed (13). Presumably, a proteinase or proteinase–inhibitor complex is activated by the sodium dodecyl sulfate treatment: The resultant active proteinase partially degrades the native 55,000-molecular-weight subunit. The protein-

Table 3.1. (*Continued*)

Inhibitor and Abbreviation	Effective Concentration (m*M*)	Comments
Serine proteinases		
Diisopropylfluoro-phosphate[c] (DFP)	1	*Very toxic*. Store 0.1 *M* stock solutions in dry isopropanol at −20°C in small aliquots. Dilute 10-fold before use.
Trasylol[d]	*e*	Nontoxic. M_r = 11,600 pI 10.5 Stable in neutral to acid media, K_i for trypsin at pH 7.8, 2×10^{-11} *M* (23).
Metalloproteinases		
Ethylenediaminetetra-acetic acid (EDTA)	1–10	Also inactivates other metalloenzymes.
o-Phenanthroline	1	Also inactivates other metalloenzymes.
Acid (carboxyl) proteinases, pH opt 2–5		
Pepstatin	0.001–0.0001 (10 μg/mL)	Reversible weak inhibitor above pH 6.0 (24).
Diazoacetylnorleucine methyl ester + Cu^{2+}	1 m*M* of each	Reacts with other proteins, particularly sulfhydryl enzymes (25).
Thiol proteinases, pH opt 4–7		
p-Hydroxymercuri-benzoate (PMB)	1	Reacts with other sulfhydryl enzymes.

[a] Many other inhibitors of microbial (26), plant (27), or synthetic (28,29) origin have been described.
[b] Also known as Dichlorvos or Vapona.
[c] All operations with pure DFP and concentrated solutions (>1 m*M*) are to be done in the hood with good air flow. Wear gloves at all times (polyvinyl gloves are recommended) and take caution not to contaminate clothing. Immediate access to atropine is strongly recommended as a precaution against accidental exposure to DFP. Aqueous DFP solutions <1 m*M* may be used outside of hood, but do not allow contact with skin (30). Place contaminated glassware in 0.5 *M* NaOH for at least 24 hr for complete hydrolysis.
[d] Trasylol is isolated from beef lung and supplied commercially. It is identical in structure to the pancreatic trypsin inhibitor (31).
[e] Use 24–50 Kallikrein units/mL (32).

ase contamination is separated from the pyruvate kinase by gel permeation chromatography on Sephadex G-100.

The following evidence shows that an enzyme purified from yeast has not been modified by proteolysis: (a) the antibody against the pure enzyme as an antigen exhibits a single fused precipitin line on Ouchterlony double diffusion against both homogeneous enzyme and enzyme in the cell free extract; (b) immunoelectrophoresis of the cell free extract produces one precipitin arc with the same mobility as pure enzyme; (c) immunoprecipitates of both the pure enzyme and of the enzyme in the cell-free extract give a single protein band in addition to the antibody on sodium dodecyl sulfate polyacrylamide gels having the same electrophoretic mobility as native enzyme; and (d) homogeneous enzyme and enzyme in the cell-free extract have the same pI (36).

Erythrocytes

After washing erythrocytes with isoosmotic buffers, the cells are lysed under hypotonic conditions (37) or by blending (38). Hemoglobin is removed from the hemolyzate by passing the extract through diethylaminoethyl- (DEAE) cellulose or by adding ethanol–chloroform (see Table 1.2). Because hemoglobin is not adsorbed to DEAE Sephadex A-50 at pH 7 (39), it can be removed if the enzyme of interest is adsorbed.

Solid Tissues

The disruption of solid tissues (muscle, liver, brain, etc.) can be accomplished with a variety of devices (4), including the Potter–Elvehjem and Dounce homogenizers, tissue presses, and blenders. More fibrous tissue such as heart and uterine muscle and skin is usually disrupted with a Polytron (4).

Myosin comprises the greatest percentage of muscle protein. Because it is not soluble at ionic strengths below 0.3 *M*, extraction with H_2O or low ionic strength buffers gives a crude extract enriched in the muscle cytoplasmic proteins. On the other hand, muscle extracts to be used for enzyme assays should be prepared at high ionic strength (0.1 *M* phosphate buffer, pH 7 plus 0.3 *M* KCl) to prevent the loss of several enzymes that associate with myofibrils at low ionic strength. This adsorption is so extensive that it forms the basis of a purification scheme for seven muscle enzymes (40).

Proteins in a crude liver extract like those in the crude yeast extract are subject to proteinase degradation. The majority of the liver proteinases are located in lysosomes and, except for the carboxyl proteinase cathepsin D, are primarily thiol proteinases (25, 41). In addition to using proteinase inhibitors described in Table 3.1, rat liver proteinase activity can be destroyed by heating at 60–65°C for 5 min (42). Liver tissue also contains phosphoprotein phosphatases. Isolating intact phosphorylated proteins from liver extracts, therefore, requires the addition of 50 m*M* sodium fluoride (43).

Plant Tissue

Many plant tissue extracts contain large amounts of phenolic compounds. If not removed immediately, these compounds or their oxidation products form insoluble protein complexes. The usual procedure is to extract plant tissue with buffer containing 2% insoluble polyvinylpyrrolidone (polyclar AT) (44). Additional fractionation steps, such as an acid precipitation (45), may also be necessary to remove the phenols. Some plant tissue extracts also contain a significant amount of nucleic acid, which should be precipitated before proceeding with the enzyme fractionation.

Membrane Enzymes

Membrane proteins are divided into two classes, peripheral or extrinsic proteins and integral or intrinsic proteins (Figure 3.2). Peripheral proteins interact with the membrane by nonpolar and electrostatic interactions and are usually dissociated from the membrane with 1–5 m*M* EDTA, 0.1–1 *M* NaCl, or 1–10% acetic acid (46). The binding of some peripheral enzymes, known as ambiquitous enzymes, is under metabolic control so that substrates or cofactors specifically dissociate them (47).

Peripheral and soluble proteins are distinguished from intrinsic membrane proteins by their inability to associate with nonionic detergent micelles. When nonionic detergent micelles associate with integral membrane proteins, they can be detected by noting the effect of ionic detergents on their electrophoretic mobility. The ionic detergent is incorporated into the protein–nonionic-detergent micelle and changes its electrophoretic mobility (48). The mobility of proteins not associated with the nonionic detergent micelle is unaffected. Another method of distinguishing peripheral membrane proteins from intrinsic membrane proteins

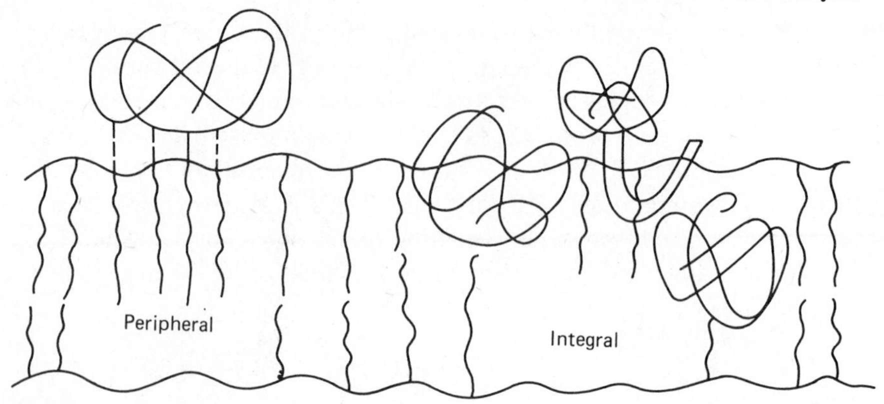

Fig. 3.2. Schematic representation of the association of peripheral and integral proteins with a phospholipid bilayer.

involves dispersing the protein in Triton X-114 at 0°C. When the temperature of this detergent is raised above 20°C, its cloud point, the solution separates into two phases, an aqueous phase and a detergent phase. Hydrophilic proteins are recovered in the aqueous phase whereas integral membrane proteins are found in the detergent phase (49).

How integral membrane proteins associate with the phospholipid bilayer (as depicted in Figure 3.2) is determined by the distribution of hydrophobic and hydrophilic residues on the protein surface. Proteins immersed in the hydrophobic region of the bilayer have nonpolar residues on their surface. Peripheral proteins are more amphiphilic and associate with the membrane to varying degrees.

Solubilizing integral membrane proteins requires detergents or organic solvents (46, 50–52). The solvating environment of the lipid bilayer must be replaced so that the protein can be purified. Integral proteins are kept in these solvents throughout their purification; in the absence of these solvents, they tend to form aggregates, which are insoluble in aqueous buffers. Because of these difficulties, few membrane proteins have been purified to homogeneity (53).

Many membrane preparations also contain proteinases (46); as with yeast and liver enzyme preparations, proteinase inhibitors (Table 3.1) should be added to membrane preparations to reduce proteinase degradation.

No one detergent or organic solvent can be identified as "most effective" for membrane solubilization (51); the best detergent must be found empirically. In most cases, integral membrane proteins are solubilized and purified in nonionic detergents (46, 54), because they are effective in breaking lipid–lipid and lipid–protein interactions, and seldom break interactions between proteins or protein subunits so that enzymic activity is preserved. Occasionally, membrane proteins form artifactual aggregates in nonionic detergents. Solubilizing these nonspecific protein complexes requires other detergents. Integral membrane enzymes that require phospholipids for activity (55, 56) require careful control of the pH and ionic strength to maintain their enzymic activity. Finally, some integral proteins buried in the lipid bilayer require an organic solvent for solubilization. Solvents most commonly used are chloroform–methanol mixtures, 75% ethanol, butanol, butanol–methanol mixtures, butanol–methanol–ammonium acetate, and pyridine (46).

Nonionic and zwitterionic detergents are recommended for membrane solubilization because they do not interfere with most fractionation methods (53). Detergents with a high critical micelle concentration are recommended when analyzing the biological function of membrane proteins because they are more readily replaced by dialysis, thus functioning more effectively in reconstitution experiments (57). Because detergents with a low critical micelle concentration are more effective in keeping proteins in their native state, they are preferred when characterizing them (52).

The properties and structures of some detergents that are useful for solubilizing and purifying membrane proteins are given in Table 3.2 and Figure 3.3. If one desires a more comprehensive compilation, refer to Helenius et al. (51). Commercial detergents, such as Triton X-100 and Tween 80, typically contain contaminants that oxidize sulfhydryl groups (62). If necessary, they can be removed (62), or, alternatively, one can add 1 mol of a free radical scavenger, such as butylated hydroxy toluene, per 500 moles of detergent (51). A sulfhydryl reducing compound can also be added to protect sulfhydryl enzymes. Detergents that contain a phenyl group, like Triton X-100, absorb UV light and thus interfere with many spectrophotometric measurements, particularly protein measurements at 280 nm. However, Triton X-100 can be hydrogenated to reduce its absorption at 280 nm without seriously affecting its physical properties (63). Finally, the heterogeneous nature of some detergents adversely affects certain of their properties. For example, high- and low-molecular-weight components in the heterogenous Triton X-100 polymer interfere in membrane reconstitution experiments (58). Fortunately, the heterogeneous Triton X-100 can be fractionated by silica gel chromatography (58): The polymer composed of 11 ethyleneoxy units functions best during the re-

Table 3.2. Useful detergents for isolating integral membrane proteins

Detergent	Critical Micelle Conc. (mM)[a]	Monomers per Micelle	Molecular wt of Micelle ($\times 10^{-3}$)	Comments
Triton X-100	0.3	140	90	Polydisperse (58). ϵ_{274} nm = 2.32 mL/mg cm (59).
Tween 80	0.012	60	76	Polydisperse, purity varies. Absorption bands at 268 and 279 nm given Tween 80 its yellow color (51).
β-D-Octyl-glucoside	25	27	8	Dialyzes rapidly. No absorption in the ultraviolet (UV) or visible region.
Lauryl maltoside	0.2	98	50	New detergent (60) with properties similar to Triton X-100. No absorption in visible or UV region.
Egg lysolecithin	0.02–0.2	181	95	Polydisperse. Subject to autoxidation.
Sodium cholate	3[b]	4.8[b]	2[b]	$pK_a = 5.2$
3-[(3-Cholamidopropyl) dimethylammonio] 1-propane sulfonate (CHAPS)	4–6			Zwitterionic ϵ_{280} nm = $3(M\ \text{cm})^{-1}$ (61)

[a] These values are approximate; the actual value depends on the temperature, ionic strengh, and pH.
[b] The critical micelle concentration is particularly sensitive to pH; at pH 8–9, 0.15 M NaCl, critical micelle concentration equals 3 mM.

constitution procedures. [Techniques and strategies for removing detergents from proteins are reviewed in Reference (64).]

The best way to purify an integral membrane protein is to selectively extract it from the membrane (51). Adding 0.1 mg detergent/mg lipid initiates the extraction process but leaves the phospholipid bilayer essentially intact. Adding 2 mg detergent/mg lipid results in the formation of soluble lipid–protein–detergent, protein–detergent, and lipid–detergent

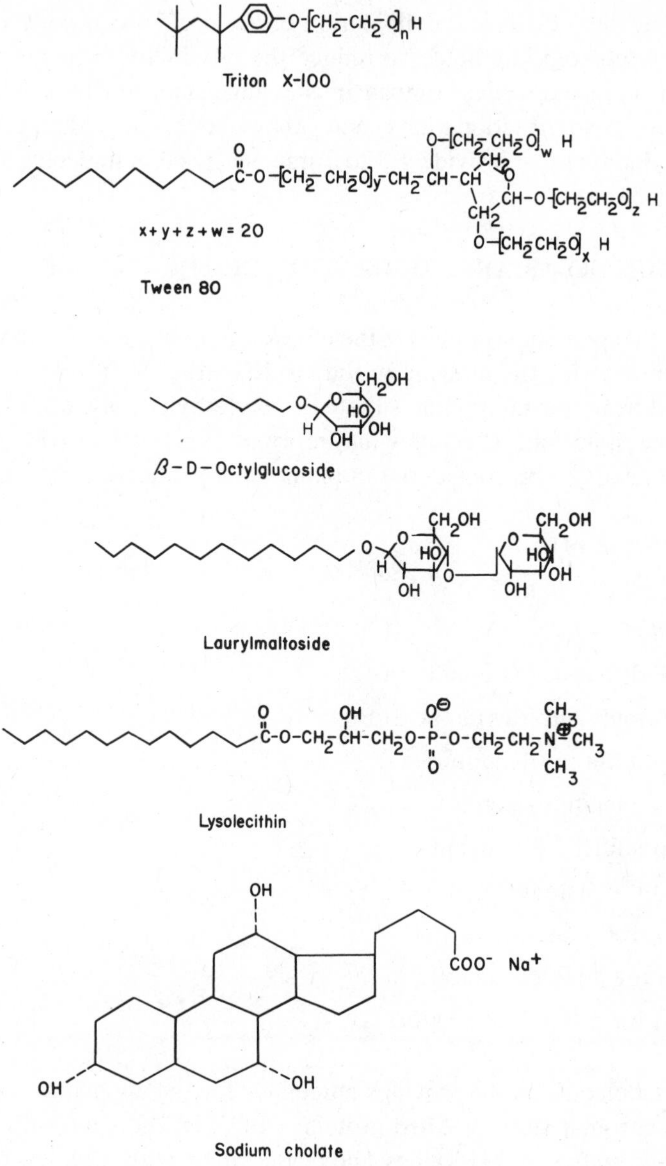

Triton X-100

$x + y + z + w = 20$

Tween 80

β-D-Octylglucoside

Laurylmaltoside

Lysolecithin

Sodium cholate

Fig. 3.3. Structures of some detergents for solubilizing membrane enzymes.

mixed micelles. Complete delipidation of the protein is usually achieved at 10 mg detergent/mg lipid. To reduce the possibility of having two different protein molecules trapped in the same micelle, the solvent used for subsequent fractionations and characterizations should contain enough detergent to provide 1.5 to 2 micelles/protein molecule (51).

PRELIMINARY CHARACTERIZATION OF THE ENZYME

The next step in our strategy for the development of a purification scheme is to characterize the enzyme in the crude extract or in the concentrate obtained from the ammonium sulfate or poly(ethylene glycol) (PEG) fractional precipitation. The following information about the enzyme to be purified needs to be considered in planning a purification scheme:

- Isoelectric pH
- Molecular weight
- Stability
  - To different pH conditions
  - To high and low temperatures
  - To different denaturants
  - To chelating agents
  - To sulfhydryl reagents
- Kinetic constants
  - K_m for substrate(s)
  - K_i for possible inhibitors
  - K_a for possible activators

The isoelectric point provides guidelines for the application of ion exchange chromatography. Most proteins can be fractionated with an anion exchange matrix at pH values above the pI or with a cation exchange matrix at pH values below the pI. The molecular weight of the native enzyme in the crude extract can be determined either by polyacrylamide gel electrophoresis or by gel permeation chromatography (see Chapter 4 for more details). If the enzyme has a very large or very small molecular weight, gel permeation chromatography is indicated as an early step in the purification because the molecular weights of most proteins range between 40,000 and 300,000. Separating any protein that falls outside of this size range should result in a significant purification, although this approach is not practical for large volumes because of the cost of gel permeation media.

Knowing the stability of the enzyme under a variety of conditions serves two purposes. First, it defines the conditions required to maintain the activity of the enzyme during purification. Second, if the enzyme shows unusual stability to pH or temperature extremes, or to organic solvents and denaturants, such conditions might be used to denature a significant proportion of the contaminating proteins in the crude extract.

Examine the stability of the enzyme under the following conditions (be sure to include proteinase inhibitors in extracts from yeast, liver, and membranes):

pH values from 2 to 10

2-Mercaptoethanol at 1, 10, and 50 m*M*

EDTA at 1 and 10 mM

Glycerol at 10 and 25%

Ethanol at 1, 2, 5, 10, and 20%

KCl or NaCl at 0.05, 0.1, and 0.2 *M*

KSCN or NaSCN at 0.3, 0.6, 1, and 2 *M*

Incubate samples both at 4°C and room temperature; assay for enzyme activity at time zero and after 24 hr. Samples that are stable at room temperature for 24 hr should be incubated for extended times at more elevated temperatures, such as 37, 45, and 55°C to test for unusual stabilities. The K_m for substrate(s), K_a for activators or cofactors, and the K_i for possible inhibitors may be useful in defining specific ligands for affinity chromatography.

Certain properties of the enzyme in the crude extract, particularly the pI, molecular weight, and K_m, K_i, and K_a values, should be compared with their respective values for the purified enzyme to determine whether they were altered during purification.

PURIFICATION OF AN ENZYME

This section describes a variety of fractionation techniques for purifying enzymes. The order of their presentation does not imply a specific importance; the intent is to provide enough detail so that informed decisions can be made regarding the sequence of steps to follow when developing a purification scheme. Additional details regarding a particular methodology can be obtained by consulting the references.

The condition(s) used throughout the purification is(are) selected on the basis of the results of the preliminary characterization. Performing the fractionation steps at room temperature is preferable if the stability study warrants it. Include glycerol in the solvent if it stabilizes the protein because it does not interfere in most fractionation methods, however, it is expensive so use a minimum concentration. If the enzyme is inactivated by EDTA or 2-mercaptoethanol, it is likely to be a zinc metalloenzyme. The loss of zinc from zinc metalloproteins may be prevented by maintaining 10^{-5} to 10^{-6} M zinc in all buffers; higher concentrations of Zn^{2+} inactivate most enzymes.

The goal of every purification scheme is to obtain a homogeneous enzyme with a high specific activity in as high a yield and in as short a time as possible. Achieving this goal requires a systematic approach to the fractionation strategy and an understanding of the theory and practice of the fractionation procedure. Furthermore, the fewer steps in the procedure the better the yield and the shorter the time required to complete it. A four step purification scheme with a 90% yield at each step gives an overall yield of 63%, which is excellent. Occasionally, a fractionation step is selected to give a higher specific activity at the expense of yield. This practice is more likely at the last step of a purification scheme, especially if a homogeneous enzyme is obtained.

Fractional Precipitation of Crude Extracts

Primarily because of proteinase contamination, crude extracts should not be stored. The enzyme should be concentrated by fractional precipitation with ammonium sulfate or PEG, or adsorbed and desorbed from a chromatographic matrix as soon as possible. If the crude extract must be stored, it is preferable to freeze it in liquid N_2 and store it at −80°C.

To determine what concentration of ammonium sulfate or PEG to use, add increasing amounts of the chosen precipitant to a series of small volumes of the extract. After equilibration and centrifugation, determine the amount of enzyme and protein remaining in the supernatant liquid of each small volume. Using these data, choose the two concentrations of precipitant to use. The precipitate obtained with the lowest concentration is discarded and that obtained with the highest concentration is retained. Deciding which concentration of precipitant to use is the first of many decisions that must be made in an enzyme purification. Fractionations are never sharp because some enzyme is discarded or lost in every step; the question is, how much can one afford to lose.

Most fractional precipitation procedures are completed at 4°C because the precipitates are usually collected by centrifugation with a refrigerated centrifuge. If precipitations are completed at room temperature, the temperature of centrifugation should be controlled accordingly. Make sure the centrifuge is engineered to run at ambient temperatures.

Precipitated proteins, after dissolution in a small volume, are more stable because they are more concentrated, and ammonium sulfate and PEG are stabilizing additives. Other stabilizing compounds, which are lost during the precipitation steps, such as enzyme cofactors and proteinase or phosphatase inhibitors, can also be replenished.

Fractionating a crude extract by salting out with ammonium sulfate is a convenient and useful purification step. The procedure can be completed by (a) adding the dry salt (pulverize salt with mortar and pestle to increase rate of dissolution), (b) adding a saturated solution, or (c) dialyzing against a concentrated solution. Add dry salt when working with large volumes; adding a saturated ammonium sulfate solution gives less surface denaturation but may produce large and unwieldy volumes. However, adding a saturated solution to precipitate protein from small volumes of a concentrated protein sample, usually encountered at later stages of the purification, is more convenient than adding the dry powder. Low concentrations of protein (0.01 to 0.1 mg/mL), often encountered in fractions from a chromatographic column, are conveniently precipitated by dialysis against saturated ammonium sulfate. Adding a saturated solution or the dry salt to a dilute protein solution increases the volume significantly, resulting in a poor recovery. More precise fractionations can be achieved with a reverse extraction procedure, that is, by extracting a protein precipitate with a graded series of decreasing concentrations of ammonium sulfate (65).

Analytical reagent grade $(NH_4)_2SO_4$ is not pure enough for most protein fractionations. Crude extracts may be fractionated with this grade of salt but purified or partially purified enzyme preparations should be fractionated with a recrystallized salt. Ammonium sulfate can be recrystallized from water containing 1 m*M* EDTA, or an enzyme grade ammonium sulfate, which is available from several commercial sources, can be used.

It is also a good practice to include 1 m*M* EDTA in all protein solutions fractionated with $(NH_4)_2SO_4$.

Because $(NH_4)_2SO_4$ hydrolyzes in water to give a pH near 5, the pH of the protein sample may need to be adjusted as ammonium sulfate is added. Adjust the pH as the salt is added or adjust the pH of the saturated $(NH_4)_2SO_4$ solution with 6 *N* NH_4OH before adding it. Since the pH of saturated $(NH_4)_2SO_4$ cannot be measured directly with a pH meter, dilute a measured volume 10 fold and adjust the pH. Calculate the amount of NH_4OH to add to the saturated ammonium sulfate based on the amount of base required to adjust the pH of the diluted sample. Using pH paper also gives more accurate pH values with dilute solutions.

The effectiveness of ammonium sulfate in precipitating a protein depends on the protein concentration, the pH of the solution, and the temperature. Less salt is needed at higher protein concentration, at a pH near the pI of the protein, and at higher temperature. Protein solubilities in $(NH_4)_2SO_4$ can also be influenced by protein aggregation. For example, a unique purification scheme for porcine liver and kidney phosphofructokinase, developed by Massey and Deal (66), takes advantage of this behavior. At low ionic strength, in the absence of Mg^{2+}, the enzyme aggregates and is precipitated by low concentrations of $(NH_4)_2SO_4$. Adding fructose 6-*P* and adenosine triphosphate (ATP) reverses the aggregation and solubilizes the enzyme.

The amount of $(NH_4)_2SO_4$ required to achieve various levels of saturation are given in Tables 3.3 to 3.6. The concentrations of ammonium sulfate listed in these tables are expressed in terms of molarity rather than percent saturation since molarity is much less a function of temperature than percent saturation (67). A word of caution is in order: If a near saturated 3.8 *M* $(NH_4)_2SO_4$ solution is used, it will absorb water from the atmosphere and change concentration if not stoppered well or prepared fresh each time it is used. This uptake of water is not a problem with saturated solutions, as long as excess solid $(NH_4)_2SO_4$ remains suspended in the solution. The final concentration of $(NH_4)_2SO_4$ after a fractionation step is easily determined by comparing its conductivity to a standard curve of conductivity versus concentration.

In principle, fractionating proteins with PEG can be as effective as fractionating with ammonium sulfate. Poly(ethylene glycol) has the general formula $HO\text{-}(CH_2\text{-}CH_2\text{-}O\text{-})_nH$ where $n > 4$. A number of polymers with differing weight-average molecular weights are available, often under

Table 3.3. Grams of ammonium sulfate to add to 1 L of solution at 0°C

Percent Saturation	Initial Molarity	Final Molarity																				
		0.00	0.20	0.40	0.60	0.80	1.00	1.20	1.40	1.60	1.80	2.00	2.20	2.40	2.60	2.80	3.00	3.20	3.40	3.60	3.80	3.90
0.0	0.00	0.00	26.7	54.0	81.9	111	140	170	202	234	267	302	338	375	413	453	495	539	585	632	682	707
5.1	0.20		0.00	27.0	54.7	83.0	112	142	173	205	238	272	308	344	383	422	464	507	552	599	649	673
10.3	0.40			0.00	27.4	55.4	84.2	114	144	176	209	243	278	314	352	391	432	475	519	566	615	639
15.4	0.60				0.00	27.7	56.2	85.5	116	147	179	213	247	283	321	359	400	442	486	533	581	605
20.5	0.80					0.00	28.1	57.1	87.0	118	150	183	217	252	289	328	368	409	453	499	546	570
25.7	1.00						0.00	28.6	58.1	88.5	120	153	186	221	258	296	335	376	420	465	512	535
30.8	1.20							0.00	29.1	59.1	90.2	122	156	190	226	264	303	343	386	430	477	499
35.9	1.40								0.00	29.6	60.2	91.9	125	159	194	231	270	310	351	395	441	464
41.1	1.60									0.00	30.2	61.4	93.7	127	162	199	236	276	317	360	405	428
46.2	1.80										0.00	30.7	62.6	95.7	130	166	203	242	282	325	369	391
51.3	2.00											0.00	31.3	63.9	97.7	133	170	208	248	290	333	355
56.5	2.20												0.00	32.0	65.2	99.8	136	174	213	254	297	318
61.6	2.40													0.00	32.7	66.7	102	139	178	218	260	281
66.8	2.60														0.00	33.4	68.2	104	142	182	224	244
71.9	2.80															0.00	34.2	69.8	107	146	187	207
77.0	3.00																0.00	35.0	71.5	110	150	169
82.2	3.20																	0.00	35.8	73.2	112	132
87.3	3.40																		0.00	36.7	75.0	94.0
92.4	3.60																			0.00	37.6	56.1
97.6	3.80																				0.00	18.1
100.0	3.90																					0.00

[a] Reprinted by permission from Reference (67).

Table 3.4. Milliliters of a 3.8 M ammonium sulfate solution to add to 1 L of solution at 0°C

Initial Molarity	Final Molarity																	
	0.00	0.20	0.40	0.60	0.80	1.00	1.20	1.40	1.60	1.80	2.00	2.20	2.40	2.60	2.80	3.00	3.20	3.40
0.00	0.00	55.3	117	185	263	351	452	570	709	875	1077	1330	1655	2088	2693	3600	5111	8134
0.20		0.00	58.4	124	197	281	377	489	621	779	972	1213	1522	1933	2508	3371	4809	7684
0.40			0.00	61.9	132	211	302	408	534	683	866	1094	1387	1777	2322	3140	4503	7228
0.60				0.00	65.9	141	227	327	446	587	760	975	1252	1620	2135	2907	4194	6768
0.80					0.00	70.5	152	246	357	490	652	855	1115	1462	1946	2673	3884	6305
1.00						0.00	75.9	164	268	393	545	735	978	1303	1756	2437	3570	5837
1.20							0.00	82.3	179	295	437	613	840	1143	1565	2199	3255	5366
1.40								0.00	89.7	197	328	492	702	981	1372	1959	2936	4891
1.60									0.00	98.7	219	369	562	819	1179	1718	2616	4412
1.80										0.00	110	247	423	657	984	1475	2294	3931
2.00											0.00	124	282	494	789	1232	1971	3447
2.20												0.00	141	330	593	988	1645	2961
2.40													0.00	165	396	742	1319	2472
2.60														0.00	198	496	991	1981
2.80															0.00	248	662	1489
3.00																0.00	332	994
3.20																	0.00	498
3.40																		0.00

[a] Reprinted by permission from Reference (67).

Table 3.5. Final volume in milliliters after addition of solid ammonium sulfate to 1 L of solution at 0°C

	Final Molarity																				
Initial Molarity	0.00	0.20	0.40	0.60	0.80	1.00	1.20	1.40	1.60	1.80	2.00	2.20	2.40	2.60	2.80	3.00	3.20	3.40	3.60	3.80	3.90
0.00	1000	1010	1021	1033	1046	1060	1074	1090	1106	1123	1142	1161	1181	1203	1225	1249	1275	1301	1329	1359	1373
0.20		1000	1011	1023	1035	1049	1063	1079	1095	1112	1130	1149	1169	1191	1213	1237	1262	1288	1316	1345	1359
0.40			1000	1012	1024	1037	1052	1067	1083	1100	1118	1137	1157	1178	1200	1223	1248	1274	1301	1330	1344
0.60				1000	1012	1025	1039	1054	1070	1087	1105	1124	1143	1164	1186	1209	1233	1259	1286	1315	1329
0.80					1000	1013	1027	1042	1057	1074	1091	1110	1129	1150	1172	1194	1218	1244	1271	1299	1313
1.00						1000	1014	1028	1044	1060	1077	1096	1115	1135	1156	1179	1203	1228	1254	1282	1296
1.20							1000	1014	1030	1046	1063	1081	1100	1120	1141	1163	1187	1211	1237	1265	1278
1.40								1000	1015	1031	1048	1065	1084	1104	1125	1147	1170	1194	1220	1247	1260
1.60									1000	1016	1032	1050	1068	1088	1108	1130	1152	1176	1202	1228	1242
1.80										1000	1016	1033	1052	1071	1091	1112	1135	1158	1183	1209	1222
2.00											1000	1017	1035	1054	1073	1094	1116	1140	1164	1190	1203
2.20												1000	1018	1036	1056	1076	1098	1122	1145	1170	1183
2.40													1000	1018	1037	1058	1079	1101	1125	1150	1162
2.60														1000	1019	1039	1060	1082	1105	1129	1142
2.80															1000	1019	1040	1062	1085	1109	1121
3.00																1000	1020	1041	1064	1087	1099
3.20																	1000	1021	1043	1066	1077
3.40																		1000	1022	1044	1055
3.60																			1000	1022	1033
3.80																				1000	1011
3.90																					1000

[a] Reprinted by permission from Reference (67).

Table 3.6. Final volume in milliliters after addition of a 3.8 M ammonium sulfate solution to 1 L of solution at 0°C

Initial Molarity	Final Molarity																	
	0.00	0.20	0.40	0.60	0.80	1.00	1.20	1.40	1.60	1.80	2.00	2.20	2.40	2.60	2.80	3.00	3.20	3.40
0.00	1000	1051	1109	1174	1248	1333	1432	1547	1683	1847	2047	2298	2621	3051	3654	4560	6070	9091
0.20		1000	1055	1117	1187	1268	1362	1471	1601	1757	1947	2186	2493	2902	3476	4337	5773	8646
0.40			1000	1059	1126	1202	1291	1394	1517	1665	1846	2072	2363	2751	3294	4111	5472	8196
0.60				1000	1063	1135	1219	1317	1433	1573	1743	1957	2232	2598	3112	3883	5168	7741
0.80					1000	1068	1147	1239	1348	1479	1640	1841	2099	2444	2927	3652	4862	7282
1.00						1000	1074	1160	1262	1385	1535	1723	1966	2289	2741	3420	4552	6818
1.20							1000	1080	1176	1290	1430	1605	1831	2131	2553	3185	4240	6350
1.40								1000	1088	1194	1324	1486	1694	1973	2363	2948	3924	5878
1.60									1000	1097	1216	1365	1557	1813	2171	2709	3606	5402
1.80										1000	1108	1244	1419	1652	1979	2469	3287	4922
2.00											1000	1122	1280	1491	1785	2227	2965	4441
2.20												1000	1141	1328	1590	1984	2642	3956
2.40													1000	1164	1394	1740	2316	3469
2.60														1000	1198	1494	1989	2979
2.80															1000	1248	1661	2488
3.00																1000	1331	1994
3.20																	1000	1498
3.40																		1000

[a] Reprinted by permission from Reference (67).

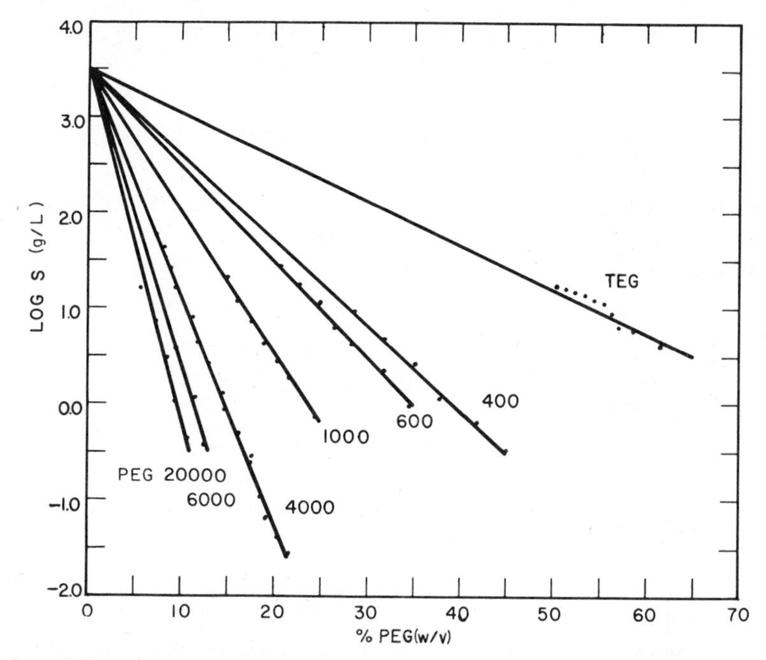

Fig. 3.4. Effect of molecular weight of poly(ethylene glycol) (PEG) on the solubility of human serum albumin. Measurements were made in 0.05 *M* potassium acetate buffer, pH 4.5, containing 0.1 *M* KCl. Total protein concentration was 20 mg/mL, except for PEG-4000 data, which ranged from 2 to 125 mg/mL. The solid lines for PEG-400, -600, -1000, and -4000 are least squares fits. The lines for PEG-6000, PEG-20,000, and TEG were arbitrarily drawn through the common intercept. TEG is $HO(CH_2—CH_2)_4H$. Reproduced by permission of the author. Reference (68).

the trade name Carbowax. They are distinguished by the shorthand notation, PEG-*XX*, where *XX* is the weight-average molecular weight.

Poly(ethylene glycol) polymers have a variety of commercial uses. In recent years they have gained popularity as effective precipitating agents in the fractionation of protein mixtures. Fractionating proteins with PEG is useful since protein solutions containing the nonionic PEG need not be dialyzed before subjecting them to a variety of other separation procedures.

Figure 3.4 shows that all PEG polymers precipitate proteins; on a (w/v) basis, smaller amounts of the higher molecular weight homologs are required. For example, precipitating 95% of a 20 mg/mL solution of human serum albumin requires 10% PEG-20,000, 12% PEG-6000, 15% PEG-4000, 24% PEG-1000, 33% PEG-600, and 39% PEG-400 (68). Enzymologists most often use PEG-4000 and PEG-6000. The other homologs, particularly the smaller ones, may be more useful for some proteins, but no practical guidelines are available because each protein behaves differently.

The effectiveness of PEG as a precipitating agent, like that for $(NH_4)_2SO_4$, is strongly influenced by conditions that affect the self-association or aggregation of proteins (69, 70). Moderate concentrations of PEG-4000 (10–20%) precipitate chymotrypsin at conditions that cause the enzyme to self-associate, namely at pH 8.5 and low ionic strength. Salt (0.1 *M*) reverses this self-association and prevents precipitation by 25% PEG-4000. Fifteen percent PEG-4000 precipitates polymeric glutamate dehydrogenase; again conditions that prevent polymer formation, 1 m*M* guanosine triphosphate (GTP) plus 1 m*M* nicotinamide adenine dinucleotide (reduced) (NADH), prevent precipitation (69). Even aggregates of different proteins are precipitated by lower concentrations of PEG than their dissociated forms (70). Finally, it takes less PEG to precipitate large proteins than small proteins (68).

Poly(ethylene glycol) can be added either as a solid or as a solution. Fifty percent solutions are often used. As with 3.8 *M* $(NH_4)_2SO_4$, 50% solutions of PEG should be stored in well sealed containers because such solutions absorb water from the atmosphere and become more dilute.

Even though PEG does not interfere with ion exchange chromatography, it does interfere with other procedures often encountered during a purification. Moderate concentrations of PEG alter the properties of Sephadex and interfere with some analytical procedures (71, 72). Four percent solutions of PEG decrease the bed volume of Sephadex G-200 by 57% and increase the partition coefficient (see Chapter 4 for a definition of partition coefficient) for albumin from 0.37 to 0.91. Poly(ethylene glycol) polymers also give artificially high values for plasma protein concentrations determined by radial immunodiffusion.

Removing PEG from protein solutions is more inconvenient than removing $(NH_4)_2SO_4$. Most PEG homologs cannot be removed by dialysis, however, they may be removed by gel permeation chromatography, ultrafiltration, or a salt-induced phase separation (71). Poly(ethylene glycol) polymers elute much earlier from gel permeation columns than globular proteins of comparable size. For example, PEG-2000 elutes in the same volume as RNase (M_r = 13,000) on a Bio-Gel P-30 column (71). Poly(ethylene glycol) forms a separate phase after adding inorganic phosphate. At 0.7 *M* potassium phosphate, pH 7, a 20% PEG-4000–protein solution separates into two phases: 99% of the protein partitions in the lower phosphate phase and nearly 99% of the PEG partitions in the upper phase (71). Proteins can also be removed from PEG solutions by ion exchange or affinity chromatography or by precipitation methods (71). Methods are available [see, e.g., (73)] to determine the amount of PEG remaining in the solution.

Denaturing Contaminating Proteins

If the enzyme to be purified shows unusual stability to extremes of pH or temperature, or to an organic solvent, take advantage of this property early in the development of the purification scheme. Select the experimental conditions that lead to the most extensive loss of contaminating proteins and a good recovery of the enzyme.

If the desired enzyme is stable to pH extremes, use weak acids or bases to adjust the pH of the protein solution. Scopes (74) suggests 1 *M* acetic acid or 5 *M* acetate buffer adjusted to a pH no more than two units lower than the desired pH. To raise the pH, use 1 *M* Tris (p*K* 8.1), 1 *M* 2-methyl-2-amino-1,3-propanediol (p*K* 8.8), or 10% NH_4OH (p*K* 9.3); stir the solution vigorously to prevent denaturation around the drops as they disperse. After removing denatured proteins by centrifugation, adjust the pH of the supernatant liquid to the desired value.

Heat denaturation is a harsh method for fractionating proteins. Unless the enzyme of interest can be stabilized by a specific cofactor or some other additive, fractionating proteins by heat denaturation is not recommended. If used in the purification scheme, it should be completed in a carefully controlled manner. Stirred solutions should be brought to temperature rapidly with a water bath set at 10°C above the desired temperature. Maintain the solution at the desired temperature for a fixed period of time and then cool it rapidly on ice. Because of the time required to heat and cool, volumes in excess of 100 mL should be divided. Crude extracts or fractions with a proteinase contamination are usually not fractionated by heat denaturation as heating simply accelerates the rate of proteolytic degradation.

Fractionating protein solutions with organic solvents may be done in one of two ways, either by fractional precipitation of the enzyme of interest using ethanol or acetone or by preferential denaturation of the contaminating proteins in the solution. Because organic solvents such as acetone or ethanol precipitate proteins by a different mechanism than $(NH_4)_2SO_4$ or PEG, it may be useful to try them (74). After dialyzing to equilibrate the ionic strength to approximately 0.01 *M*, adjust the protein concentration to about 20 mg/mL then place the protein solution in an alcohol ice bath and stir continuously, maintaining the temperature at 0 to −5°C as the cold solvent is added to it. Ice crystals in the protein solution formed during the early phases of the solvent addition may be advantageous in maintaining the temperature. As the organic solvent concentration increases, the temperature of the solution should be lowered accordingly; for example, at 20% organic solvent, the temperature should be near −20°C. The precipitates should be recovered by centrifuging with a refrigerated centrifuge maintained at the temperature of the precipitating fraction.

Because of an ever increasing number of more effective fractionation techniques and because of the technical difficulties associated with it, fractionating protein solutions with organic solvents is seldom used. On the other hand, preferential denaturation of contaminating proteins may be more useful, as indicated by a recent purification of yeast alcohol dehydrogenase utilizing the unusual stability of this enzyme in ethanol (75).

Gel Permeation Chromatography

Gel permeation or gel filtration chromatography separates proteins by size. It is usually used in the final stages of a purification when volumes are small to remove small amounts of contaminating proteins. Unusually large ($M_r > 300,000$) or small ($M_r < 30,000$) proteins may be successfully fractionated from the crude extract using gel permeation chromatography. β-galactosidase ($M_r = 520,000$) is purified 10 fold from a crude extract with a 90% yield by high pressure gel permeation chromatography (76).

A variety of gel permeation media are available. Many are listed in Table 3.7 along with some performance criteria for each. Sephadex, a trade name of Pharmacia Fine Chemicals, is a cross-linked dextran produced by the fermentation of glucose. Several types of Sephadex with differing porosities are available. The identification numbers correspond approximately to the water regain of the dried gel multiplied by five. Sephacryl is the trade name for dextran gel cross-linked by *N*,*N*-methylene-bisacrylamide. Because it is more rigid than Sephadex, higher hydrodynamic pressures can be used with Sephacryl and sharper separations can be obtained with it by working with a finer mesh size. Both Sephadex and Sephacryl contain free carboxyl groups (4–8 μeq/g), consequently buffers should contain at least 0.02 *M* salt to prevent adsorption of proteins to the matrix. Both matrices also have an affinity for aromatic groups and for the metallocations, Cu^{2+} and Fe^{3+}.

Polyacrylamide gel filtration media are available under the trade name Bio-Gel from Bio-Rad Laboratories. The identifying numbers on the several different types indicate an approximate value for the exclusion limit of the gel for globular proteins given in 1000 molecular mass units. Bio-Gel has the same performance characteristics as Sephadex including about 5 μeq of free carboxyl groups per gram of dry gel. When compared to Sephadex G-75 superfine and Ultrogel AcA54, Bio-Gel P-100 (minus 400 mesh) has the optimal resolving power for proteins in the molecular weight range between 13,000 and 68,000. Using this gel, it is possible to completely resolve proteins differing in molecular weight by as little as 5000 (77).

Agarose gels suitable for gel permeation chromatography are available from several commercial firms. Provided in the wet state, they cannot be dried and conveniently reswollen. The numbers on the labels of the different types of Sepharose denote the percentage of dry gel in the particles. The numbers for the Bio-Gel A matrices indicate the approximate exclusion limits for proteins in million molecular mass units. Different grades of agarose with different melting temperatures (50 to 65°C) are also available. Agarose should not be used outside the pH range 4.5 to 9.

Sepharose CL (cross-linked), obtained by cross- linking Sepharose with 2,3-dibromopropanol, has a greater thermal and chemical stability than Sepharose; Sepharose CL can be autoclaved without losing its chromatographic properties. Ultrogel is prepared by polymerizing polyacrylamide inside the beads of agarose gel. The two components are physically interspaced with each other, not joined chemically; the resulting gel has good mechanical stability. Size resolution of macromolecules during column chromatography with this gel is limited by flow rates rather than by mechanical strength.

Fractogel TSK is a synthetic hydrophilic vinyl gel permeation material composed exclusively of C, H, and O atoms. This gel has a high mechanical stability even in the large pore sizes, is highly resistant to microbial decomposition, and is chemically stable from pH 1 to pH 14. Protein denaturants such as urea, guanidine·HCl, and sodium dodecyl sulfate have little influence on its performance.

Porous glass beads are available (not listed in Table 3.7) but they are seldom used in gel permeation chromatography of proteins because they

Table 3.7. Technical data for commercial gel filtration media[a]

Gel Filtration Media	Particle Size	Fractionation Range of Globular Proteins	Swelling Time (hr)		Maximum Operating Pressure (cm H_2O)	Flow Rate (mL/hr)[b]	Bed Volume (mL/g)
			20°C	90°C			
Sephadex							
G-10	Medium	700	3	1	*c*	*c*	2–3
G-15	Medium	1,500	3	1	*c*	*c*	3–4
G-25	Coarse	1,000–5,000	3	1	*c*	*c*	4–6
	Medium	1,000–5,000	3	1	*c*	*c*	
	Fine	1,000–5,000	3	1	*c*	*c*	
	Superfine	1,000–5,000	3	1	*c*	*c*	
G-50	Coarse	1,500–30,000	3	1	*c*	*c*	9–11
	Medium	1,500–30,000	3	1	*c*	*c*	
	Fine	1,500–30,000	3	1	*c*	*c*	
	Superfine	1,500–30,000	3	1	*c*	*c*	
G-75	Medium	3,000–80,000	24	3	160	77	12–15
	Superfine	3,000–70,000	24	3	160	18	
G-100	Medium	4,000–150,000	72	5	96	50	15–20
	Superfine	4,000–100,000	72	5	96	12	
A-50 m	Coarse	$100{,}000–50 \times 10^6$	Preswollen		50	30	
	Medium					10	
A-150 m	Coarse	$10^6 - 150 \times 10^6$	Preswollen		30	15	
	Medium					4	
Bio-Gel polyacrylamide							
P-2		100–1,800	4	1	*c*	*c*	3.5
P-4		800–4,000	4	1	*c*	*c*	5
P-6		1,000–6,000	4	1	*c*	*c*	8

P-10		1,500–20,000		1	*c*	*c*	9
P-30		2,500–40,000	4	1	*c*	*c*	11
P-60	Coarse	3,000–60,000	4	1	100	95	14
	Medium				100	30	
	Fine				100	...	
P-100	Coarse	5,000–100,000	4	1	100		15
	Medium				100		
	Fine				100		
P-150	Coarse	15,000–150,000	4	1	100	45	18
	Medium				100	25	
	Fine				100	...	
P-200	Coarse	30,000–200,000	4	1	75	22	25
	Medium				75	11	
	Fine				75	...	
P-300	Coarse	60,000–400,000	4	1	60	15	30
	Medium				60	6	
	Fine				60	...	
G-150	Medium	5,000–300,000	72	5	36	23	23–30
	Superfine	5,000–150,000	72	5	36	6	18–22
G-200	Medium	5,000–600,000	72	5	16	12	30–40
	Superfine	5,000–250,000	72	5	16	3	20–25
S-200	Superfine	5,000– 250,000	Preswollen		300	30	
S-300	Superfine	10,000–1.5 $\times$ 10^6	Preswollen		300	25	
Sepharose							
2B	Medium	70,000–40 $\times$ 10^6	Preswollen		30	10	
4B	Medium	60,000–20 $\times$ 10^6	Preswollen		60	12	
6B	Medium	10,000–4 $\times$ 10^6	Preswollen		90	14	
CL 2B	Medium	70,000–40 $\times$ 10^6	Preswollen		50	15	
CL 4B	Medium	60,000–20 $\times$ 10^6	Preswollen		120	26	
CL 6B	Medium	10,000–4 $\times$ 10^6	Preswollen		120	30	

Table 3.7. *(Continued)*

Gel Filtration Media	Particle Size	Fractionation Range of Globular Proteins	Swelling Time (hr)		Maximum Operating Pressure (cm H_2O)	Flow Rate (mL/hr)[b]	Bed Volume (mL/g)
			20°C	90°C			
Bio-Gel agarose							
A-0.5 m	Coarse	10,000–500,000	Preswollen		100	110	
	Medium					35	
	Fine					15	
A-1.5 m	Coarse	$10{,}000–1.5 \times 10^6$	Preswollen		100	90	
	Medium					30	
	Fine					10	
A-5 m	Coarse	$10{,}000–5 \times 10^6$	Preswollen		100	70	
	Medium					20	
	Fine					9	
A-15 m	Coarse	$40{,}000–15 \times 10^6$	Preswollen		90	50	
	Medium					15	
	Fine					6	
Ultrogel							
Ac A 54	Medium	6,000–70,000	Preswollen		*d*	3–6	
Ac A 44	Medium	12,000–130,000	Preswollen		*d*	3–6	
Ac A 34	Medium	20,000–400,000	Preswollen		*d*	3–6	
Ac A 22	Medium	$60{,}000–1 \times 10^6$	Preswollen		*d*	2–3	
Fractogel TSK							
HW 40 F	Fine	100–10,000	Preswollen		*e*	81	
HW 40 S	Superfine	100–10,000	Preswollen		*e*	*f*	
HW 55 F	Fine	1,000–700,000	Preswollen		*e*	74	

HW 55 S	Superfine	1,000–700,000	Preswollen	*e*	*f*
HW 65 F	Fine	50,000–1 $\times$ 10^6	Preswollen	*e*	*f*
HW 75 F	Fine	500,000–50 $\times$ 10^6	Preswollen	*e*	*f*

[a] The data are from the manufacturers' technical literature. Columns 2.5 $\times$ 30 cm were used for measurements on materials from Pharmacia Fine Chemicals; for the materials from Bio-Rad, Columns 1.1 $\times$ 13 cm were used. Linear flow rates of columns with larger diameter are larger than columns with smaller diameter. Operation at 4°C decreases the flow rate.
[b] Flow rate is for each square centimeter of cross-sectional area.
[c] Flow rate and hydrostatic head is limited by the efficiency of resolution. The highest resolution is obtained when the flow rate is maintained in the range of 2–10 mL/cm^2 hr.
[d] Ultrogel beads are less compressible, permitting higher flow rates than other comparable gel media.
[e] This media will withstand pressures up to 100 psi.
[f] The flow rates depend on the packing velocities; the slower the packing velocity the slower the flow rate. Columns should not be eluted at flow rates that exceed the packing velocity. These flow values are obtained with columns 1.6 $\times$ 60 cm packed at rates near 75 mL/hr cm^2.

adsorb proteins both electrostatically and by nonpolar interactions. Silica based matrices for high pressure gel permeation chromatography discussed later are also available (76, 78). For a more detailed discussion of the structure, properties, application, and availability of gel permeation media, consult Reference (79) or relevant commercial literature.

Gel permeation columns, particularly those prepared from natural materials, should be stored with antimicrobial agents. Antimicrobial compounds that have been used are 2.5% I_2 in half-saturated KI (80), 0.01–0.02% chlorotone [Cl_3C—$C(OH)(CH_3)_2$], 0.02% NaN_3 and 0.002% Hibitane (use the more soluble crystalline diacetate of Hibitane, solubility 1.9%) (79). Solutions may also be cold sterilized with diethylpyrocarbonate (81). Toluene, phenol, cresol, formalin, and $CHCl_3$ have been used, but these compounds are not recommended because they are not only toxic but some may denature proteins, and some may dissolve plastics. Sodium azide, NaN_3, should be used with caution since it is mutagenic and a suspected carcinogen.

Ion Exchange Chromatography

Separating proteins using ion exchange chromatography continues to be an important and popular technique for purifying proteins (82). Conditions can usually be established so that most proteins are adsorbed to an ion exchange medium. The pH of the protein solution is adjusted above or below the pI and the ionic strength is reduced to allow adsorption to either an anion or a cation exchanger. Proteins can usually be eluted with good yields by 0.1 to 1 *M* concentrations of salt. Because the adsorption of protein to an ion exchange media is not selective, ion exchange chromatography is usually not suitable for purifying an enzyme that constitutes only a small fraction of the total protein.

The structures and p*K* values of the functional groups of some common ion exchangers, coupled to cellulose, Sephadex, or agarose are given in Figure 3.5 It is important to know that the exchange capacity of all ion exchange media is not the same. Cellulose media vary from one supplier to another and Sephadex and agarose ion exchangers have higher capacities than their cellulose counterparts. The Sepharose CL and Bio-Gel A exchangers do not swell or shrink with changes in pH or ionic strength, as the cellulose exchangers do, and they also have a more uniform size range and are supplied prewashed.

Structure	Name		pK
$-OCH_2COO^-$	Carboxymethyl -	(CM)	~4
$-O-\overset{O}{\overset{\parallel}{P}}-O^-$ (O^-)	Phospho-	(P)	<2~6
$-OCH_2CH_2SO_3^-$	Sulphoethyl-	(SE)	<0
$-OCH_2CH_2CH_2SO_3^-$	Sulphopropyl-	(SP)	<0
$-OCH_2CH_2\overset{\oplus}{N}H(CH_2CH_3)_2$	Diethylamino-ethyl-	(DEAE)	~9.5
$-OCH_2CH_2\overset{\oplus}{N}(CH_2CH_3)_3$	Triethylamino-ethyl-	(TEAE)	

Fig. 3.5. Structure and p*K* values of the functional groups of some common ion exchangers for the chromatography of proteins.

The DEAE group is used primarily for neutral and acidic proteins. Because it loses its charge above pH 9, DEAE cannot be used for binding proteins at high pH. Above pH 9, the quaternary aminoethyl or TEAE exchanger is required. Neutral and basic proteins are usually chromatographed on carboxymethyl cellulose. Binding proteins at pH values below 3 requires the sulfoethyl or sulfopropyl group. The behavior of phosphocellulose is more complicated; its charge changes from one to two as the pH is raised from 5 to 9.

In many separations, phosphocellulose functions as a bio-ligand affinity column. For example, adenosine monophosphate (AMP) aminohydrolase is adsorbed to cellulose phosphate directly from a crude extract (83). Potassium chloride in excess of 0.7 *M* is required for elution; the protein can also be eluted with 0.025 *M* sodium pyrophosphate, a strong inhibitor (38). Also, several enzymes from baker's yeast having affinities for phosphoryl groups are successively purified by adsorption onto phosphocellulose followed by specific elution with one of their substrates (84).

The desorption of proteins from an ion exchange matrix by ampholytes, either with or without a gradient, has recently been described as ampholyte displacement chromatography. Because ampholytes are expensive, they are most useful for separating isozymes with different pI values, particularly during the final stages of a purification, when smaller volumes are involved (85–87).

In 1977, Sluyterman and Elgersma (88) described another method known as chromatofocusing, which separates proteins on the basis of their pI values. It is particularly useful for separating isozymes with different pI values. The column is first equilibrated at a high pH before the protein is applied. It is then eluted with a special amphoteric buffer, which generates a pH gradient over the desired pH range. The protein migrates down the column until it encounters a pH equivalent to its pI. When it moves further down the column, it becomes negatively charged and binds to the column; back diffusion is restricted because the proteins become positively charged and repel the matrix. Consequently, the proteins move down the column as focused bands and elute in order of their pI values (88, 89).

Unless otherwise indicated by technical literature from the commercial source, cellulose ion exchangers should be washed before use. Although some of the current commercial ion exchangers do not have to be washed, it may be necessary to remove small particles (fines) to permit good flow rate. To remove fines, suspend the medium in a large volume of distilled water and leave it to settle. Then after the usable portion of the medium settles, pour off the supernatant liquid containing the fine particles. Repeat the cycle until the supernatant liquid is free of fines.

After the fines are removed, wash the chromatographic medium with two cycles of 0.5 *M* NaOH, water, 0.5 *M* HCl, and water in a sintered glass or Buchner funnel. The cation exchangers should be washed with an additional volume of NaOH to leave the chromatographic medium in the salt form, because this form is more expanded and easier to work with.

Using a pH meter, adjust the pH of the washed medium suspended in distilled water by adding an acid or base while stirring. Because of the ion exchange capacity of the chromatographic medium, the final pH is achieved slowly. Suction off the water and wash the semidry medium two or three times with the equilibration buffer before finally resuspending it in the equilibration buffer and pouring it into a column. Check the pH of the eluent after eluting 1 to 2 bed volumes of equilibration buffer to insure that the pH of the column is properly adjusted.

To establish conditions for adsorption of the protein onto and desorption from an ion exchange chromatographic medium, the protein must first be equilibrated, usually by dialysis against a buffer at low ionic strength, $I = 0.025\ M$, and at a pH about one unit above or below its pI, depending on whether an anion or cation exchanger is to be used. Apply about 1 mg of protein at a concentration of 10 mg/mL to a test column (5 × 25 mm) and wash with 2 to 3 bed volumes of the dialysis buffer. Then elute the column with 2 to 3 bed volumes each of increasing concentrations of a neutral salt such as 0.25, 0.5, 0.75, and 1 *M* NaCl or KCl in the dialysis buffer. Test whether a biospecific ligand will elute the enzyme of interest because many enzymes are specifically eluted from an ion exchange matrix by one of their substrates or effectors (90–92). This procedure, called affinity elution chromatography, takes advantage of the ability of substrates or other ligands to change the binding of an enzyme to a general ion exchange matrix (90). Additional features of enzyme–ligand interactions and how they affect the elution of proteins from an ion exchange medium provided by Scopes (93) makes it easier to apply this method.

If the protein is not adsorbed to the matrix under the original test conditions, examine the eluent for loss of contaminating proteins. In some cases, it is advantageous to remove contaminants under conditions where the enzyme of interest is not adsorbed. The protein that passes through the first test column may be applied directly without dialysis to a second test column of a different chromatographic medium. If the enzyme is adsorbed to the second column, the desorption conditions can then be investigated as above.

The chromatographic set up should include a column with a bed volume of 1 cm^3 for every 10 mg of protein. Add the previously equilibrated protein to the column at 10 mg/mL concentration. If consistent with the preliminary experiment, increase the ionic strength of the protein solution before applying it to the column to just allow adsorption. This procedure increases the capacity of the column for the enzyme of interest by preventing adsorption of some of the contaminating protein. Generally, it is convenient to elute the protein with a linear gradient having a total volume about five times the bed volume (74).

If the protein adsorbs strongly to the chromatographic medium, that is, a high concentration of salt is required to elute it, it may be profitable to adsorb it directly from the crude extract, the ammonium sulfate, or the PEG fraction in a batchwise procedure. Add the chromatographic medium to the protein solution, stir it gently for at least 1 hr, and allow it to settle. After the medium has settled, pour off the supernatant liquid, wash the medium on a sintered glass filter with the protein solvent, pour the medium containing the protein into a column, and elute with a linear or stepwise gradient. In some cases, it may require extended periods of time for the enzyme to be adsorbed to the chromatographic medium, that is, it may be necessary to allow the suspended medium to equilibrate with the protein solution overnight.

Hydrophobic Chromatography

As noted in Chapter 1, an appreciable percentage of the surface residues of a protein molecule in aqueous solution are nonpolar. This nonpolar property is utilized by enzymologists during hydrophobic chromatography as exemplified by Hofstee's (94) observation that 11 out of 13 arbitrarily chosen proteins were adsorbed to an agarose matrix containing covalently attached hydrophobic residues.

Interactions between nonpolar groups are promoted by increased temperature, increased concentration of structure stabilizing salts (e.g., sodium phosphate and ammonium sulfate), and solvents such as glycerol and PEG. Such interactions are weakened by decreased temperature, decreased concentration of $(NH_4)_2SO_4$, or sodium phosphate, increased concentration of chaotropic salts such as NaSCN, guanidinium thiocyanate, or urea, and by solvents containing ethylene glycol (above 30%) or detergents. As one would expect then, these compounds or conditions are utilized during the adsorption or elution of proteins from hydrophobic columns (95).

Hydrophobic chromatographic media are prepared by coupling nonpolar groups to agarose or Sepharose. At least three different procedures are in use. The first procedure involves the reaction of a hydrophobic amine with cyanogen bromide (96) or the much less toxic *p*-nitrophenyl cyanate (97) activated agarose. The second method involves the reaction of the glycidyl ether of a hydrophobic alcohol with agarose (98). Product A (Figure 3.6) is obtained from the glycidyl ether method and product B from the cyanogen bromide method. The third method gives product C. It is obtained from the reaction of a hydrophobic amine with agarose activated with *p*-toluenesulfonyl chloride (99) or *p*-trifluromethylbenzenesulfonyl chloride (100). Chromatographic media prepared with the latter two activating agents are more stable than those prepared with cyanogen bromide. Sepharose can also be oxidized with bromine (101) or periodate (102) and amines coupled to the treated Sepharose by reductive amination.

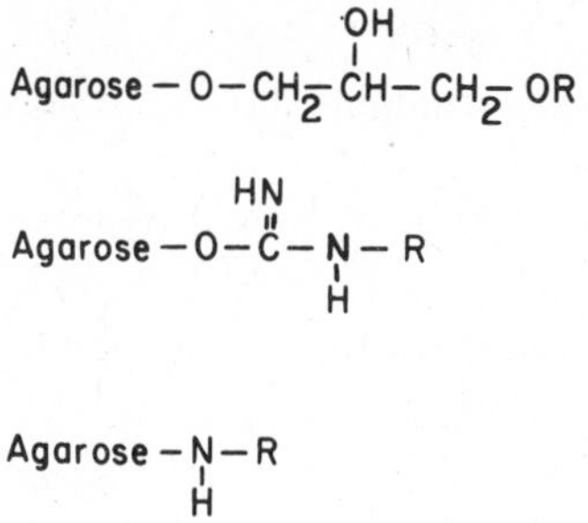

$$\text{Agarose}-O-CH_2-\underset{}{\overset{OH}{\overset{|}{C}H}}-CH_2-OR \quad (a)$$

$$\text{Agarose}-O-\overset{HN}{\overset{\|}{C}}-\underset{H}{\underset{|}{N}}-R \quad (b)$$

$$\text{Agarose}-\underset{H}{\underset{|}{N}}-R \quad (c)$$

Fig. 3.6. The covalent linkages obtained after coupling a ligand R to agarose activated with glycidyl ether (*a*), cyanogen bromide (*b*), and tosyl or tresyl chloride (*c*).

The products are given by structure C in Figure 3.6. Hydrophobic media containing a wide range of different nonpolar groups are available.

Since the isourea linkage obtained by the cyanogen bromide coupling procedure is protonated under normal chromatographic conditions (see Figure 3.6), there is some debate as to whether a medium prepared by this procedure is hydrophobic. Evidence presented by Hofstee (103) indicates that binding depends on the cooperation between hydrophobic and electrostatic forces. Wilchek and Miron (104) note that α-lactalbumin and ovalbumin bind tightly to alkylamino agarose with hydrocarbon chains of four to six carbon atoms; acetylation of the alkylisourea linkage, however, destroys the binding. On the other hand, evidence from Halperin et al. (105) shows that alkyl agarose columns with equal ligand density prepared by either the glycidyl ether or the cyanogen bromide method have essentially identical chromatographic properties. Adsorption and desorption of proteins from these columns behaved as predicted by hydrophobic interactions.

The first reported hydrophobic chromatographic separations involved the induced adsorption of protein to a variety of matrices by 1.5 to 3 *M* $(NH_4)_2SO_4$ or 1 *M* potassium phosphate, followed by desorption with a decreasing salt gradient (106, 107). The terms salting out (108) or salting in chromatography (109) are sometimes used to describe this type of protein fractionation. Some common matrices are DEAE cellulose, celite, Sepharose 4B, and L-valyl agarose (107). Proteins also adsorb to celite in the presence of PEG-6000 and desorb during a decreasing concentration gradient of this solvent (110).

Fujita et al. (111) describe the elution of proteins bound to cellulose in 2.4 to 3 *M* $(NH_4)_2SO_4$ by an increasing concentration gradient of ethanol (0.75–1.5 *M*), sucrose (0.6–1.5 *M*), glycerol (1–1.5 *M*), or urea (2–5 *M*). These authors called this method hydrogen bond chromatography because the hydroxyl group concentration in a cellulose bed is high (est 3.7 *M*), and all the eluents are expected to effectively compete for hydrogen bonds. As these compounds also affect nonpolar interactions, the observations of Fujita et al. (111) are included under the heading of hydrophobic chromatography.

Interest in the use of hydrophobic chromatography increased significantly following the emergence of a report by Shaltiel and Er-El (112) showing the effectiveness of a homologous series of ω-alkyl agaroses (Sepharose—N—$(CH_2)_n$—NH_2) in separating proteins. Glycogen synthetase is not adsorbed to the C_2 and C_3 homologs when equilibrated in 50 m*M* glycerophosphate, pH 7.5; yet it is adsorbed to the C_4 homolog and eluted with an increasing concentration gradient of NaCl. Higher members of this homologous series bind the synthetase so tightly that it cannot be eluted with a salt gradient. Glycogen phosphorylase, on the other hand, is not bound to agarose having side chains shorter than five carbons. It is possible, therefore, to vary and select the chromatographic medium to bind a specific enzyme. In fact, two types of test kits containing five small columns with Sepharose—NH—$(CH_2)_n$—NH_2 or Sepharose —NH—$(CH_2)_nH$ (where n = 0, 2, 4, 6, or 8) are commercially available. After testing the adsorptive and desorptive behavior of the protein of interest, additional amounts of the medium can be synthesized or procured commercially to chromatograph the protein. Horiuti and Imamura (113) used this approach to purify an extracellular lipase. After synthesizing a series of fatty acid esters of cellulose differing in hydrocarbon length from C_2 to C_{18}, they found the C_{16} homolog to be the most effective in adsorbing the enzyme.

Amberlite (G-50), a weak acidic cation exchange resin containing carboxyl groups, also functions as a hydrophobic chromatographic medium when protonated at pH 4.0. Several pancreatic enzymes adsorb to the resin and are effectively eluted by increasing the pH to 6. Changing the pH then changes the hydrophobicity of the medium (114).

One of the advantages of using hydrophobic chromatography is that a variety of eluting regimens function effectively. Proteins that bind to a nonpolar matrix in high salt are desorbed by a decreasing salt gradient or by a gradient of 0–50% ethylene glycol imposed on a decreasing salt gradient of 2 to 0 *M* NaCl (115). Increasing salt gradients or gradients containing 0–50% ethylene glycol (95), 0–70% propylene glycol (116), or excess Triton X-100 (116), glycerol, *N,N*-dimethyl-formamide, or sucrose (117) are also effective eluting systems. Lowering the temperature assists

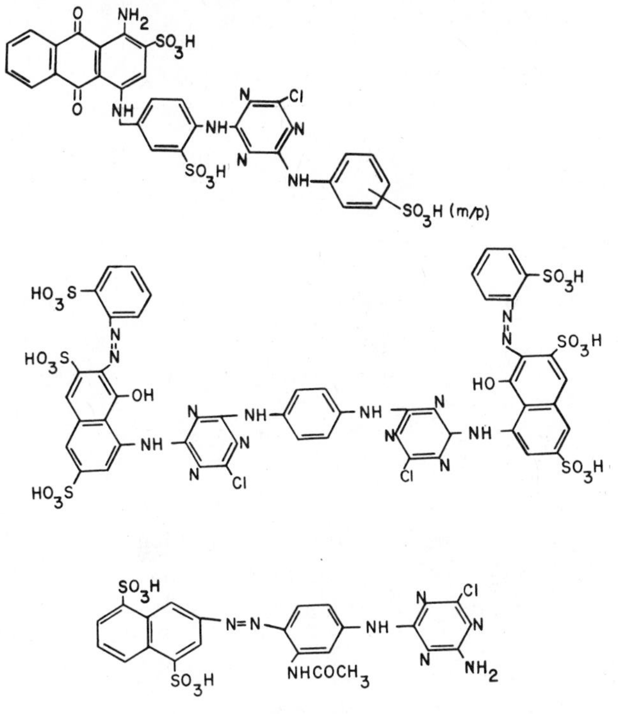

Cibacron blue F3G-A
Procion blue H-B
C.I. 61211
(Blue A)

Procion red HE-3B
(Red A)

Cibacron yellow RA
Procion yellow H-A
C.I. 13245
(Orange A)

Fig. 3.7. Structures of three typical triazine dyes. The dyes are identified by their commercial names, Cibacron (Ciba–Geigy) and Procion (ICI, Imperial Chemistry Industries), by their color index constitution number (CI), and by trivial names in parenthesis.

in eluting proteins from a hydrophobic medium by decreasing nonpolar interactions. Chaotropic compounds such as urea and NaSCN are also effective eluents but usually at the expense of protein stability. Predicting the behavior of different salts in eluting proteins from a nonpolar matrix (118) is discussed in the next section.

Dye-Ligand Affinity Chromatography

Dye-ligand affinity chromatography is useful because certain dyes interact strongly with anionic binding sites of some enzymes (119, 120). The structures of three of the dyes, which are available covalently coupled to agarose, are shown in Figure 3.7; some properties of these and other dyes are listed in Table 3.8. Dyes of this type, being flexible molecules, can

Table 3.8. Properties of triazine dyes useful in dye-ligand affinity chromatography[a]

Dye	Color Index	Mol Wt	Absorbance Max (nm)	Absorbance Coefficient (M^{-1} cm^{-1})
Procion blue MYR	61,205	636	600	4,100
Cibacron blue F3GY	61,211	774	615	7,500
Procion red H8BN		801	546	21,388
Procion yellow H-A	13,245	578	384	8,900
Procion green H-AG		1,760	676	56,400
Procion brown MX-5BR		588	530	7,780

[a] Adopted by permission from D. A. P. Small, T. Atkinson, and C. R. Low, J. Chromatogr., **216,** 175–190 (1981). Reference (121).

orient their aromatic, polar, and anionic groups to complement a wide variety of anionic enzyme binding sites (121, 122). Yet, the dyes also show specificity of binding because they do not interact strongly with all enzymes.

Of all the matrices to which the triazine dyes have been covalently attached (119, 122, 123), agarose is the most desirable. Spacer arms, described in the section on bio-ligand affinity chromatography, are not used to couple these dyes to the agarose matrix, because they are detrimental to the chromatographic process. However, tetraiodofluorescein, a smaller dye than the triazine dyes, requires a nine-atom spacer arm to function effectively (124). The dyes shown in Figure 3.6 are easily immobilized by adding agarose to an aqueous solution of the dye followed by addition of solid NaCl to give a 2% (w/v) final concentration. After the dye is physically adsorbed to the matrix, add Na_2CO_3 or NaOH to increase the pH to 10.5. Reaction will begin immediately and should be allowed to proceed for 24 to 120 hr (123). Coupling occurs when the chloro group(s) of the dye has(have) been displaced by the agarose.

As indicated in several reviews of the application of dye-ligand affinity chromatography to the purification of proteins (119, 121–123), many points are worth noting. A wide variety of enzymes with differing specificities can be purified using this method. Dye-ligand affinity chromatography is particularly effective for the purification of enzymes that bind nucleotides, including cyclic nucleotides, flavin nucleotides, and polynucleotides. Included among this group of enzymes are aminoacyltransfer ribonucleic acid (*t*RNA) synthetases, restriction endonucleases, deoxyribonucleic acid (DNA) polymerases, RNA polymerases, polynucleotide kinase, and T4 DNA ligase (125). The dye matrices show binding specificity, for example, the Cibacron blue F3G-A dye has a higher affinity for the nicotinamide adenine dinucleotide (NAD^+) dependent dehydrogenases than the Procion red HE-3B dye, which displays a greater affinity for the nicotinamide adenine dinucleotide phosphate ($NADP^+$) dependent dehydrogenases. In addition, of the 32 immobilized Procion dyes that were examined by Bruton and Atkinson (126), one or more but not all bind to 13 aminoacyl-*t*RNA synthetases.

Proteins are eluted from dye-ligand affinity columns by bio-ligands or by an increasing salt gradient. Dehydrogenases and kinases can be selectively eluted with NADH or NADPH and adenosine triphosphate (ATP), respectively; inorganic phosphate is the eluent of choice for the aminoacyl-*t*RNA synthetases (126). The adsorption of T4 DNA ligase requires a high salt concentration so it elutes in a decreasing salt gradient (125). Proteins in nonionic detergents may require the addition of an ionic detergent for them to bind. The ionic detergent forms mixed anionic micelles with the nonionic detergent which, because of charge interaction, have a weaker affinity for the dye ligand and thus allows the protein to effectively compete with the nonionic detergent for the negatively charged dye ligand (127). Lactate dehydrogenase in 2% Triton X-100, for example, has a weak affinity for immobilized Cibacron blue F3G-A; adding 1% (w/v) final concentration deoxycholate facilitates both the quantitative retention of the enzyme and its biospecific elution by 1 m*M* NADH. Sodium dodecyl sulfate (0.5% w/v) facilitates the adsorption of cyclic nucleotide phosphodiesterase from a 2% Triton X-100 crude brain extract (127).

Not all salts are equally effective in eluting protein from a dye-ligand affinity column. For instance, the difference between the concentrations of KSCN and LiCl required to elute lactate dehydrogenase from blue dextran-Sepharose, called the elution concentration, is 475 m*M*. The elution concentration is the salt concentration at the midpoint of an enzyme elution profile obtained by an increasing salt gradient. For the blue dextran-Sepharose columns, it correlates with few exceptions to the viscosity *B* coefficients of the Jones–Dole equation (128). The larger the *B* coefficient of an anion or cation, the greater its tendency to structure water and promote nonpolar interactions, such as those involved in the interaction of proteins with a hydrophobic chromatographic support medium. Therefore, the higher the elution coefficient of the salt, the higher the

Table 3.9. Properties of several salts and their effectiveness in eluting lactate dehydrogenase from blue dextran-Sepharose[a]

Salt	Identification Number	Elution Conc (m*M*)	*B* Coefficient	Elution Coefficient
$BaCl_2$	1	105	0.206	−0.124
KSCN	2	120	−0.110	−0.110
KI	3	180	−0.075	−0.075
KBr	4	250	−0.039	−0.039
CsCl	5	260	−0.052	−0.052
$CaCl_2$	6	260	0.271	−0.045
KNO_3	7	275	−0.053	−0.037
RbCl	8	300	−0.037	−0.037
$MgCl_2$	9	300	0.371	−0.025
$(CH_3CH_2)_4NCl$	10	300	0.374	−0.025
KCl	11	330	−0.014	−0.014
K phosphate buffer, pH 7.5	12	330		−0.009
NH_4Cl	13	430	−0.014	0.041
NaCl	14	500	0.079	0.079
Tris–HCl	15	510	0.343	0.081
CH_3CO_2K	16	525	0.243	0.089
$(NH_4)_2SO_4$	17	540	0.194	0.096
KF	18	570	0.093	0.093
LiCL	19	595	0.142	0.123

[a] Adapted by permission from Reference (118).

concentration required to elute a protein bound to a column principally by nonpolar interactions.

A careful examination of the correlation between the viscosity *B* coefficients and the elution concentrations given in Table 3.9 shows that the *B* coefficients of the salts of divalent metals, ammonium, acetate, and tetraethylammonium ions do not correlate well. Such divalent metal cations as Mg^{2+} and Ba^{2+} appear to interact directly with the protein or dye and to promote elution. Why the *B* coefficients of the other salts do not correlate is not known, but to overcome this deficiency, Robinson et al. (118) developed an elution coefficient for each anion and cation listed in Table 3.10. The elution coefficient (EC) of a salt is the algebraic sum of

Table 3.10. Ion coefficients[a]

Ion	Coefficient at 25°C		dB/dT
	B	Elution	
Ba^{2+}	0.22	−0.11	
SCN^-	−0.103	−0.103	
I^-	−0.068	−0.068	0.0017
Cs^+	−0.045	−0.038	
Br^-	−0.032	−0.032	0.0011
Ca^{2+}	0.285	−0.031	
Rb^+	−0.030	−0.030	
NO_3^-	−0.046	−0.030	
$(CH_3CH_2)_4N^+$	0.381	−0.018	−0.011
Mg^{2+}	0.385	−0.011	0.0024
K^+	−0.007	−0.007	0.0012
Cl^-	−0.007	−0.007	0.0012
SO_4^{2-}	0.208	0.001	−0.0019
NH_4^+	−0.007	0.048	0.0005
Na^+	0.086	0.086	0.0000
$CH_3CO_2^-$	0.250	0.096	
F^-	0.10	0.10	
Li^+	0.150	0.132	−0.0011

[a] Adapted by permission from Reference (118).

the ion coefficients (Table 3.10); for K_2SO_4, EC $= [2(-0.007) + 0.001] = -0.013$

To predict the concentration of any salt required to elute a specific protein, determine the elution concentrations of two salts and plot them versus their known elution coefficients, as in Figure 3.8. The elution concentration of other salts with known elution coefficients can be determined directly from such a plot. To chromatographically separate a protein from a crude extract, a PEG, or an ammonium sulfate fraction, Robinson et al. (118) recommend eluting first with KCl, which has an elution coefficient near zero. A plot of the elution concentrations for the enzyme of interest and the major contaminating proteins versus the elution coefficients of KCl and another salt allows one to choose a salt that provides

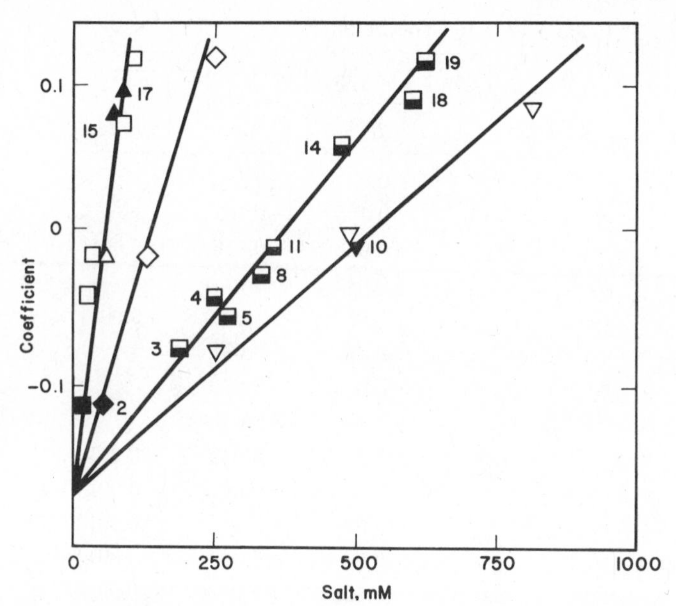

Fig. 3.8. Elution concentration of various neutral salts required to elute different proteins from blue dextran Sepharose plotted as a function of viscosity *B* or elution coefficients. (⬓) Lactate dehydrogenase; (□, ■), alcohol dehydrogenase; (△, ▲), serum albumin; (♦, ◇), pyruvate kinase; (▽, ▼), nucleoside diphosphate kinase. Open symbols (□, ▽, ⬓, △, ◇) represent measurements with simple salts plotted as *B* coefficients. Filled symbols (■, ▼, ▲, ♦) represent measurements with complex salts plotted as elution coefficients. The numbers identify the salt as indexed in Table 3.9. Reprinted by permission of the author. Reference (118).

the best separation. Salts deleterious to protein function should, of course, be avoided.

Having a large number of triazine dyes available makes dye-ligand affinity chromatography particularly attractive for enzyme purifications. As demonstrated by Scopes (129), using two different dye-ligand affinity columns in tandem—the first with a low affinity for the enzyme of interest and the second with a high affinity—it is possible, when eluting with a bio-ligand, to purify 2-keto-3-deoxy-6-phosphogluconate aldolase from *Zymomonas mobilis* to near homogeneity in one step. Of the 47 dyes examined by Scopes, only Procion yellow MX-GR binds the enzyme in 20 m*M* phosphate buffer. Therefore, Scope's procedure demonstrates again that dye-ligands show specificites for binding proteins and, in addition, that the crude extract of a biological tissue can be applied directly to the column matrix without prior removal of nucleic acids or other nonproteinaceous material.

Bio-Ligand Affinity Chromatography

Bio-ligand affinity chromatography is defined as the separation of proteins by adsorption and desorption from a chromatographic matrix containing a covalently bound biological molecule (bio-ligand). This method has revolutionized enzyme purification in many different ways (130–133). For example, proteins can be separated on columns containing their respective immobilized substrates, inhibitors, cofactors, or allosteric effectors; antigens with antibodies (134–138); glycoproteins with lectins (132, 133); haptoglobins with hemoglobins (134); repressor proteins with DNA (130); immunoglobulins with Staphalococcol protein A (139); and biotin labeled protein with avidin (140).

The success of bio-ligand affinity chromatography depends on the hydrophillic matrix, the type of derivative and spacer arm used to immobilize the ligand, and the conditions for adsorption and desorption of the protein (130, 132). Bio-ligand affinity chromatography is commonly applied in two different ways. For the first approach, one immobilizes a specific ligand that displays an absolute and unique specificity for the protein of interest. For the second approach, to be discussed in more detail later, a ligand that binds to a general class of enzymes is immobilized. Experience has shown that the loss of adsorption specificity with the second procedure is adequately compensated for by the proper choice of eluting conditions.

Of the variety of matrices currently available, agarose or cross-linked agarose is recommended. The latter is preferred when strong denaturants are used during elution. Cellulose is limited by its fibrous and nonuniform character, which impedes penetration of large protein molecules. The activation of Sephadex leads to a considerable degree of additional cross-linking, which decreases its porosity, making it relatively ineffective as a support. Bio-Gel P also appears to be limited by its low degree of porosity (130). A modified glass, glyceryl-controlled pore glass (CPG), contains a hydrophilic nonionic coating and is useful primarily in high pressure liquid chromatographic systems (78).

The effective adsorption of a protein to a bio-ligand affinity column requires an accessible bio-ligand. Ligands immobilized on a low porosity gel, such as cellulose, Sephadex, or Bio-Gel P, are not available simply because they are immobilized within the matrix and proteins are sterically excluded from interacting with them. Agaroses allowing inclusion of proteins with $M_r = 4$ to 20×10^6 do not have this problem. The uneveness in the microenvironment of an immobilized ligand can be overcome by locating the ligand at the end of a long chain or spacer arm covalently attached to the matrix.

The length of the spacer arm required for an effective matrix depends on the ligand as well as the matrix (130). Generally, for agarose, a spacer arm of four to six methylene groups between the ligand and the matrix is needed to achieve optimal interaction with the complementary protein. If proteins are used as bio-ligands, they are normally immobilized without a spacer arm.

The bio-ligand should exhibit a specific and unique affinity for the protein to be purified. For the binding of enzymes catalyzing multisubstrate reactions or enzymes activated or inhibited by allosteric effectors, a broader choice of ligands is available. Assume that ATP or NAD^+, which are substrates for a wide range of multisubstrate enzymes, is immobilized; if the other substrate of the enzyme enhances its affinity for the immobilized bio-ligand, then these effectors may be used to induce both adsorption and desorption of the enzyme. For enzyme in which the kinetic mechanism is ordered, the immobilized group specific bio-ligand may either bind first or require the second substrate for binding. If the binding requires the second substrate, its removal will induce desorption of the enzyme.

To be effective in affinity chromatography, the bio-ligand must have a reasonable affinity for the protein. If one assumes that the binding of the protein to an immobilized ligand is described by

$$\mathrm{E} + \mathrm{L} \rightleftharpoons \mathrm{EL}$$

with a dissociation constant K_D, then it can be shown by a series of rearrangemerts and assumptions (130), that

$$[\mathrm{EL}]/[\mathrm{E}_T] = [\mathrm{L}_T]/(K_\mathrm{D} + [\mathrm{L}_T])$$

where $[\mathrm{E}_T]$ and $[\mathrm{L}_T]$ are the initial concentrations of enzyme and immobilized ligand, respectively. Lowe (130) reports that if the concentration of the immobilized ligand is 5 mM, the dissociation constant for the immobilized ligand must not be larger than 0.1 mM in order to effect near quantitative adsorption of the enzyme. If K_D is larger than 0.1 mM, the concentration of the immobilized ligand should be larger. However, there is a practical limit to the concentration of ligand that can be immobilized. High concentrations of immobilized ionic or hydrophobic ligands lead to nonspecific adsorption of proteins, which destroys the selectivity of the process.

The chemistry for the activation of the matrix support and the coupling of the ligand will not be discussed. References (130–132) and commercial technical literature provide these details.

Do not become discouraged if the protein of interest fails to bind to a bio-ligand affinity column during the first try. Be prepared to vary the pH, temperature, ionic strength, and protein concentration; in short, all parameters that affect the protein–ligand equilibrium, including the time allowed for complex formation. Increasing the concentration of the protein will also favor adsorption to the matrix.

Because increased temperatures decrease the interaction of a ligand with a protein, one normally applies the protein to a column at 4°C; the protein may then be eluted at room temperature. If the bio-ligand is charged, maintain an ionic strength near 0.15 M to reduce nonspecific electrostatic interactions. Nonspecific nonpolar interactions are reduced by low concentrations of an organic solvent such as ethylene glycol.

One parameter of the interaction that is often overlooked is the rate of formation of the enzyme immobilized ligand complex. The lack of movement of the immobilized ligand and the restricted mobility of the protein within the matrix severely reduces the rate of complex formation. To overcome this problem, allow the protein solution to flow slowly over the matrix. If the volume of the protein solution is no larger than that of the matrix bed, allow the protein solution to enter the matrix, stop the flow, and incubate the matrix with the protein solution for a minimum of 1 hr. If the volume of the protein solution is considerably larger than the volume of the matrix, one may obtain adsorption in a batch procedure. Batch procedures are carried out by incubating a suspension of the affinity matrix with the protein solution while gently mixing until adsorption occurs. Twenty-four hour incubations may be necessary for dilute proteins. After the protein is adsorbed, the support medium can be removed by suction filtration, washed, and poured into a column for elution. Do not allow the bed to dry out during the suction filtration.

In most cases, protein is eluted from a bio-ligand affinity column with an increasing concentration gradient of salt, substrate, or effector. For expensive substrates, it may be convenient to use a pulse elution, in which the eluent is applied to the column in a small volume and washed through with the equilibrating buffer. The effectiveness of the salt for eluting proteins from an affinity column involving significant nonpolar interactions may be predicted from the elution coefficients developed by Robinson et al. (118), given in Table 3.10. A change in pH or an increased temperature may also affect elution (130).

When the dissociation constant of the enzyme-immobilized bio-ligand complex is very small, more stringent conditions for elution are needed. Bio-ligands in this category are antibodies, nonpolar hormones, certain vitamins, and protein receptor molecules. Several effective conditions for the desorption of proteins that have a high affinity for the immobilized ligand are listed in Table 3.11. Most of the conditions on this list were developed for eluting antigens from their immobilized antibodies. An antibody can also be modified chemically to reduce the affinity of the antigen (141). The most desirable eluent is the one that leads to the highest recovery of functional protein. Remember, it is important to determine whether or not the protein of interest is stable in the eluting solvent before beginning the elution.

High Performance Liquid Chromatography (HPLC)

Researchers in separation science have known for many years that the finer the mesh of a matrix, the higher the resolution of substances being separated on a chromatographic column. Unfortunately, it is also true that the finer the matrix, the slower the flow rate, making the time required to complete the separation prohibitive. To get around this slow flow rate, pumps have been developed and used to force the eluent (the mobile phase) through the chromatographic matrix (the stationary phase). Pumping the mobile phase creates yet another problem, namely high pressure, which in turn requires a mechanically stable stationary phase lest the matrix collapses and packs. Solving these problems led to the development of high performance liquid chromatography, often called high pressure liquid chromatography.

The application of this technology to the separation of proteins led to an extensive amount of literature (142), which indicated that an ideal packing should be (a) mechanically stable to mobile phase velocities of at least 1 mm/sec (b) completely hydrophillic (c) of high capacity, (d) chem-

Table 3.11. Typical eluents or techniques for eluting proteins that have a high affinity for an immobilized bio-ligand on a chromatographic matrix

	Reference
Acids	
0.01–1 *M* HCl	134
0.1–1 *M* acetic acid, pH 2.0	134
1 *M* propionic acid	134
0.1 *M* glycine·HCl, pH 2.2 to 2.8	134
0.1 *M* glycine·HCl, pH 2.5:0.5 *M* NaCl:1 m*M* $ZnCl_2$	135
0.015 *M* acetic acid:0.15 *M* NaCl	130
1.17 *M* NaCl:HCl, pH 2	130
Bases	
0.01 *M* NH_4OH	134
0.2 *M* 1,4-diaminobutane	130
0.1 *M* glycine·NaOH: 0.5 *M* NaCl, pH 10 to 11	138
0.025 *M* Na_2CO_3:0.5 *M* NaCl, pH 10 to 11.5	138
0.5 *M* diethylamine, pH 11.5:0.5% sodium deoxy cholate	137
Salts	
2–4 *M* KSCN or NaSCN	132,134
2.5 *M* NaI	130
2–4 *M* $MgCl_2$	134
0.08 *M* lithium diiodosalicylate	139
2 *M* sodium trichloroacetate	134
1–6 *M* guanidine·HCl	134
Neutral organics	
1–6 *M* urea	132
Electrophoretic elution	132

ically stable over a broad pH range (e) available in 5- to 10-μ particle sizes, (f) available with pore diameters from 300 to 1000 Å, (g) spherical, (h) easy to pack, and (i) inexpensive. Because no one chromatographic matrix with the above criteria is available at the present time, considerable time and effort is being expended today to develop new materials that do. Tables 3.12 and 3.13 are lists of currently available commercial matrices for high performance ion exchange and gel permeation chromatography of proteins. Procedures for preparing other matrices containing immobilized dye-ligands (121, 144), substrates (145), antibodies (145), and hydrophobic groups (146) are available.

Because most commercial columns for high performance liquid chromatography of proteins use silica as a stationary phase, the mobile phase must have an upper limit pH = 8 to prevent its dissolution. Silica matrices

Table 3.12. High performance ion exchange supports for proteins[a]

Name (Manufacturer)	Support Material	Bonded Phase	Particle Diameter (μm)	Pore Diameter (Å)	Ion Exchange Capacity	Type of Exchanger
Pharmacia (Pharmacia Fine Chemicals)						
Mono Q	Organic	Quaternary amine	10	NA[b]	0.3[c]	SAX[d]
Mono S		Sulfonic acid	10	NA	0.15[c]	SCX[d]
Polyanion Si	Silica	Polyamine	7,16	340	1[c]	WAX[d]
SynChropak (SynChrom, Inc.)						
AX 100	Silica	Polyamine	5,10	100	64[e]	WAX
AX 300			5,10	300	93	
AX 500			5,10	500	59	
AX 1000			5,10	1,000	—	
CM 300	Silica	Carboxymethyl	5,10	300	NA	WCX[d]
Toyo Soda (Toyo Soda Corp.)						
IEX 540 DEAE Sil	Silica	NA	5	130	0.3[f]	WAX
IEX 545 DEAE Sil	Silica	NA	10	240	>0.3[f]	
IEX 530 CM Sil	Silica	NA	5	130	>0.3[f]	WCX
IEX 535 CM Sil	Silica	NA	10	240	>0.3[f]	WCX

[a] Reprinted by permission from *Am. Lab.*, 14, 10, (1982) p. 30. Copyright 1982 by International Scientific Communications, Inc. Reference (143).
[b] NA signifies not available.
[c] meq/m L
[d] WAX and WCX designate weak anion and cation exchangers, respectively, while SAX and SCX represent the corresponding strong anion and cation exchanger.
[e] Ion exchange capacity is expressed in milligrams of bovine serum albumin bound per gram of support.
[f] meq/g

withstand a broad range of organic and aqueous mobile phases so the pH 8 limit is not restrictive. Of course, the mobile phase must be compatible with good recovery of enzyme activity.

The resolution and loading capacity of the stationary phase is closely controlled by its pore diameter. If the pores are small enough to exclude proteins, the capacity of the column will be severely reduced because most of the binding groups are buried in the matrix. Furthermore, when the molecular size of the protein approaches the pore size of the matrix, molecular diffusion within the pores is restricted. This restriction to diffusion also applies to mass transfer between the stationary and mobile phases, resulting in a loss of resolution. According to Regnier (142), 250- to 300- Å diameter pore packings have the broadest utility because they offer high loading capacity and good resolution of proteins with M_r = 50,000 to 100,000.

Contrary to what might be assumed, resolution on a 5-cm-long column is almost as good as that obtained with a 25-cm-long column. Consequently, it may be better to use small columns especially for analytical work because (a) smaller elution volumes are needed to elute proteins, (b) sensitivity is increased up to six times because eluents are more concentrated, (c) operating pressures are lower and (d) these columns last longer and are cheaper. The major drawback to using small columns is their lower loading capacity.

Because an increasing mobile phase velocity or pumping rate has a negative impact on the equilibrium penetration of proteins into the porous matrix, good resolution will not be obtained when linear velocities exceed 0.1 mm/sec. Columns of 0.41-cm diameter give best resolution at 0.25 mL/min; in most cases, however, flow rates of 0.5 to 1 mL/min provide adequate resolution.

High performance liquid chromatographic systems are more expensive to set up than conventional systems. High pressure liguid chromatography columns require a uniformly sized matrix and special equipment for packing them. Stainless steel columns are required to withstand the high pressures required to force the mobile phase through the stationary phase. In addition, injection ports and special pumps are required. Post column monitoring equipment, while not essential, makes the process more efficient. Yet the benefits of HPLC often outweigh the costs. One of the major benefits is speed. If only 15 min is required to complete a chromatographic separation, several separations can be completed in a day to overcome the limited loading capacity. Furthermore, sophisticated

Table 3.13. Aqueous gel permeation matrices[a]

Name (Manufacturer)	Support Material	Bonded Phase	Particle Diameter (μ*M*)	Pore Diameter (Å)	Exclusion Limit Globular Protein
Aquapore-OH (Brownlee Labs)	Spherical silica	Glycerol type	10	100	1×10^5
			10	300	1×10^4
			10	500	—[b]
			10	1000	—
			10	4000	—
CPG Glycophase (Pierce)	Controlled pore glass	Glycerol type	37–74	40	
				100	
				240	
				500	
Lichrosorb Diol (E. Merck)	Irregular silica	Glycerol type	10	100	8×10^4
Lichrospher Diol (E. Merck)	Spherical silica	Glycerol type	10	100	
			10	300	
			10	500	
			10	1000	
			10	4000	
Protein Column (Waters)					
1-60	Spherical silica	Glycerol type	10	60	2×10^4
1-125	Irregular silica		10	125	8×10^4
1-250	Irregular silica		10	250	5×10^5
SynchroPak GPC (Synchrom Inc.)	Spherical silica	Glycerol type	10	100	5×10^5
			10	300	
			10	500	
			10	1000	
			10	4000	

TSK SW (Toyo Soda Corp.)	Spherical silica	Proprietary:			
2000		contains	10	130	1×10^5
3000		hydroxyl	10	240	5×10^5
4000		groups	13	450	7×10^6(est)
Spheron (Lachema)	Spherical glycol				
P40	methacrylate		10–20	40	
P100	copolymer		10–20	100	
P300			10–20	300	
P1000			10–20	1000	
P100,000			10–20	—	
Shodex OHpak (Showa Denko)	Spherical methacrylate	10	—	—	
B-804	glycerol copolymer				
TSK PW (Toyo Soda)	Hydroxylated				
1000	polyether	10	—		
2000		10	50		
3000		13	200		
4000		13	500		
5000		17	1000		
6000		25	—		

[a] Reprinted by permission from *Am. Lab.*, **14,** 10, (1982) p. 35. Copyright 1982 by International Scientific Communications, Inc. Reference (143).
[b] Dash indicates data not available.

computer controlled gradient makers provide for the development of complex gradients that may allow the separation of complex mixtures not possible with conventional chromatographic techniques.

Fast-Flow Liquid Chromatography

Monodisperse ion exchangers give enzymologists another tool for fractionating proteins. Because most particles of this ion exchanger have the same diameter, 40% of the prepacked columns are void volume so they have a low back pressure and a fast flow rate (147, 148). The monodisperse chromatographic media are used in the same way as conventional ion exchangers except flow rates are 25–100 times faster and dilution of protein in the peak tubes is 10 times less. Because the pores of the matrix are large (exclusion limit for proteins with $M_r = 10^7$), the matrices have a large binding capacity. They also exhibit little nonspecific binding of proteins.

Three types of matrices are currently available: Mono Q is a strong anion exchanger [$—CH_2—N^+(CH_3)_3$]; Mono S is a strong cation exchanger ($—CH_2—SO_3^-$); and Mono P is used for chromatofocusing. For optimum resolution of proteins, Bergstrom et al. (148) recommend specific buffers at each pH range for each of the exchange matrices. By using buffers with pK_a values as close as possible to the pH at which the separation is achieved, pH fluctuations in the eluent can be diminished and separations improved. Sometimes separations are improved by using buffer gradients rather than salt gradients. Because there is usually a rise in pH when anion exchangers are eluted with a salt gradient, use a buffer with a pK_a slightly above the elution pH. The reverse applies when cation exchangers are used: The buffer should have a pK_a slightly below the working pH.

Electrophoretic Separation Methods

Purifying proteins by electrophoretic methods continues to be popular among enzymologists, particularly for the isolation of microgram quantities. Polyacrylamide gel electrophoresis and isoelectric focusing are the most common electrophoretic methods in use because they allow the isolation of sufficient protein for kinetic and physical characterization. Isoelectric focusing is particularly useful for eliminating microheterogeneity in proteins. For example, proteins purified in this way give a higher quality crystal for X-ray crystallographic studies (149).

Polyacrylamide is the most common support matrix for the electrophoretic separation of proteins or protein subunits (150, 151). However, other support media, such as Bio-Gel P-300 (152), can also be used. During analytical polyacrylamide gel electrophoresis, 10 to 200 μg of protein are typically separated (see Chapter 4). The fractionation of larger quantities (up to 250 mg) by preparative polyacrylamide gel electrophoresis requires special equipment (150). After completing the electrophoretic fractionation, proteins are eluted from the gel by diffusion from a gel slice after maceration in buffer or by electrophoretic methods (151). Eluting native proteins from a macerated gel slice is not recommended since recoveries are poor. However, denatured and Coomassie Brilliant Blue stained proteins are recovered in good yield by extracting a macerated polyacrylamide gel with 66% acetic acid at 4°C. Coomassie Brilliant Blue is then separated from protein by chromatography on DEAE cellulose in 66% acetic acid (153). The protein may be renatured, used for production of antisera, or analyzed in other ways.

Excellent recoveries of protein from sodium dodecyl sulfate polyacrylamide gels are also obtained by extracting with 0.1% (w/v) detergent (154). [The protein may first be visualized in the gel by KCl staining (154) or by phosphorescence (155).] To obtain native enzyme from the sodium dodecyl sulfate extract, the detergent is first removed by extracting with acetone as an ion pair. The resulting protein precipitate is dissolved in guanidine·HCl, and renatured (see Chapter 1). Alternatively, the recovery of proteins from sodium dodecyl sulfate gels can be improved if the gel is first modified by placing an intermediate layer of agarose in it. Stopping the electrophoresis when the protein is in the agarose layer allows the quantitative extraction of the protein by centrifugation. Fluorescamine labeled protein in an adjacent channel can be used to monitor its migration during electrophoresis (156).

It is important to know that proteins eluted from polyacrylamide gels may contain substantial amounts of polyacrylate (157). This material, which absorbs at 220 nm, is separated by gel permeation on Sephadex G-50, (158) or by ion exchange chromatography (157).

Desorbing the protein from a gel slice by electrophoresis, either into a small dialysis sack (159) or into a small volume on the top of a tube gel (160) gives a high yield. Pour a normal tube gel to 20 mm from the top and place the gel slice containing protein on the gel. Then reverse the poles on the electrophoresis chamber and complete the electrophoresis. A small dialysis sack may be placed over the top of the gel tube but this is not necessary if care is taken while overlaying the electrode buffer. Recoveries of 90–100% are reported. When large protein samples (250 mg) are separated by electrophoresis, use a slab gel (151). In this case electrophoresis is continued until the protein migrates out of the gel into a special collection device connected to a fraction collector.

Several experimental systems are described in the literature for purifying proteins by electrofocusing using ampholines (161) or buffer systems of defined composition (162). Sephadex G-75 (superfine) (163), low melting agarose specially treated for electrofocusing (164), Pevikon [a copolymer of polyvinyl chloride and polyvinyl acetate (165)], and glycerol (166) are used as support media. Pevikon and agarose matrices give the best results. The tendency for some proteins to precipitate at their isoelectric point may be overcome by adding a nonionic detergent, Brij 35 (0.5% final concentration) (167). Using buffers of defined composition to establish a pH gradient eliminates the problem of separating proteins from the ampholines (162) after the electrophoresis is completed. As an alternative to isoelectric focusing, an electrophoretic fractionation at a pH equal to the pI of the enzyme of interest may have advantages because the contaminants migrate leaving the enzyme where the sample was applied (168). A more extensive discussion of electrophoretic fractionation methods may be found elsewhere (169).

Miscellaneous Fractionation Methods

When solutions of the appropriate concentrations of PEG and dextran in water containing salts and buffers are mixed, two phases are formed. The top phase is rich in PEG, while the lower one is rich in dextran (170). This partitioning system, or a modification of it, forms the basis of the phase partition methods for fractionating protein. Partitioning proteins between the two phases of PEG and dextran depends on their molecular weight, charge, and hydrophobic nature (171), as well as the kinds of ions present and the ratio between different ions. The ionic strength is not so important. Adding detergent that binds to the protein amplifies the hydrophobic effect. Detergent micelles with hydrophilic groups similar to PEG, such as Triton X-100, prefer the PEG phase. Other detergents, such as digitonin, which has a carbohydrate hypdrophile, prefer the dextran phase. Hence, hydrophobic proteins, which have a tendency to associate with a Triton micelle, partition with the detergent (171). The partitioning may also be directed by covalently attaching hydrophobic groups to PEG. Palmitoyl–PEG selectively extracts serum albumin from plasma into the upper phase while almost all the remaining proteins partition in the lower phase. This method can also be used to determine the hydrophobicity of a protein (171).

The covalent attachment of a bio-ligand (172) or dye-ligand (173, 174) to PEG or dextran provides yet another important variation of the phase partition methodology. Proteins that form complexes with the ligand are partitioned into the phase carrying the ligand. In addition to partitioning the PEG-immobilized ligand between two phases, the protein–PEG–ligand complexes may be separated from other proteins by gel permeation chromatography or by ultrafiltration (172). Alternatively, if PEG and the dye- or bio-ligand are coupled to Sepharose or Sephadex, the adsorbed protein may be recovered by placing the matrix containing the bound enzyme in a column and eluting the enzyme from it (174).

Hydroxylapatite, an inorganic crystalline matrix form of calcium phosphate, is a classical support media for the chromatography of proteins. It actually preceeded the use of celluose ion exchangers. However, because the mechanisms involved in separating proteins on this media are not well defined and because many commercial preparations vary considerably, it is no longer used extensively. Sources of variability and improved methods of preparation are discussed in Reference (175).

Fractionating some sulfhydryl enzymes by covalent chromatography may be a useful procedure. In principle, a sulfhydryl compound is first immobilized and subsequently oxidized to a disulfide. Proteins bind to this support matrix by a thiol–disulfide interchange reaction (176) and are then eluted with low molecular weight thiols (Table 1.4).

Some proteins, particularly serum proteins, are successfully fractionated by chromatography on immobilized chelates of Cu^{2+} and Zn^{2+}. Acetate and phosphate buffers facilitate adsorption while amines cause elution. Chelating agents such as EDTA are also effective eluents (177).

Crystallization

Crystallizing a protein is not only aesthetically pleasing, but it may also be effective as a final step in a purification scheme (178). Many enzymes are easy to crystallize; precipitating a nearly homogeneous enzyme simply by adding $(NH_4)_2SO_4$ or PEG, often gives crystals. Some proteins, how-

Table 3.14. Basic points to consider in defining the sequence of steps in a protein purification scheme[a]

1. Take advantage of the biochemical properties of the enzyme.
2. Use a technique that is as selective as possible as soon as possible.
3. Select a high capacity technique as an initial chromatographic step.
4. Each succeeding step should utilize a different mode of separation.
5. Link chromatographic steps so that the sample obtained from one step is suitable for application in the next step.
6. Use procedures that concentrate proteins after those that dilute them.
7. Keep potential for scale up in mind.
8. Monitor the progress of purification after each step.

[a] Reprinted with permission from G. Sofer and V. J. Britton, *Biotechniques*, **1**, 198–203 (1983). Reference (184).

ever, are difficult to crystallize. The pH, protein concentration, or concentration of the precipitating agent may need to be systematically varied to achieve a crystalline protein (179–183). Because crystalline enzymes are seldom used or required for experiments, additional details for crystallizing enzymes will not be discussed.

Sequence of Fractionating Steps

The basic outline of a protein purification scheme has already been presented in Figure 3.1. The first steps in this scheme, namely, preparation of a crude extract, preliminary fractionation of the crude extract, as well as the preliminary characterization of the enzyme in the crude or concentrated extract are typical of most purification schemes. The problem now is to define the sequence of a minimum number of steps to purify the enzyme to homogeneity.

Table 3.14 contains an outline of the basic points that should be considered when developing a protein purification scheme. Item 1, how to take advantage of the properties of an enzyme when developing a purification scheme, was discussed earlier in this chapter. Item 2 suggests that the first steps in the purification should be as selective as possible. Many enzymologists, therefore, add an affinity matrix directly to the crude extract instead of initiating a fractional precipitation procedure with ammonium sulfate or PEG as was common in the recent past. After the protein of interest is adsorbed to the affinity matrix, the enzyme–matrix complex is allowed to settle under gravity or removed by suction filtration.

It is then washed by suction filtration before it is packed into a column and eluted. As called for in Item 3, the matrix used for the initial chromatographic step should have a high capacity so that the volume of the matrix required to bind all of the enzyme remains as small as possible. The smaller the matrix volume, the smaller the elution volume, and the easier it is to perform subsequent chromatographic steps.

Item 4 recommends a different mode of separation for each succeeding step in the purification. In this way, the unusual stabilities, charge, hydrophobic characteristics, size, and ligand specificities of the protein can be exploited. Furthermore, it is often easier to link one chromatographic step directly to another (Item 5), when different modes of separation are used. For example, the eluent of a gel permeation column can be applied directly to an ion exchange column and that from an ion exchange column can usually be applied directly to a hydrophobic matrix. The remaining items, Items 6, 7, and 8 on Table 3.14 apply to the development of the overall purification scheme. Item 6 emphasizes the need to use procedures that concentrate proteins after those that dilute them. The loss of protein from dilute solutions is always much greater than from concentrated solutions. Scaling up a preparation, Item 7, may or may not be important depending on the projected use of the purified enzyme. Finally, monitoring the progress of a purification scheme, as called for in Item 8, by determining both the yield of enzyme activity and protein at each step, is critical for the development of an efficient procedure. Without this information, it is impossible to intelligently select the next step to try in the development of the purification scheme.

The purification scheme developed by Scopes (129) for 2-keto-3-deoxy-6-phosphogluconate aldolase from *Z. mobilis* provides an exemplary demonstration of many of the items listed in Table 3.14. After examining a large number of dye-ligands, Scopes selected two dyes, one did not bind the enzyme but did bind many contaminating proteins in the crude extract and a second that had a high affinity for the enzyme. By linking these two dye-ligand columns in series and eluting the second column with a bio-ligand, he obtained a homogeneous enzyme in essentially one step. The dye-ligand matrices have a high capacity: 470 mg of the 800 mg in the crude extract was retained on the first column. After washing the second column with extraction buffer, homogeneous enzyme was eluted with a bio-ligand with an 83% recovery of activity. This differential dye-ligand chromatographic step exemplifies two different modes of separation. Furthermore by eluting with a bio-ligand, the second column had the highest resolving power.

PURIFICATION RECORDS

As discussed in the previous section, records of a purification scheme are kept to assist in its development and to provide guidelines for others who may wish to use it. Figure 3.9 provides a sample enzyme purification table. Once a scheme is developed, the published procedure should be clearly described. Indicate the temperature, concentration in molarity of all compounds, column dimensions, the volume of all fractions, and the concentration of protein in each fraction. Also indicate the methods used to determine protein concentration and units of enzyme activity.

STORING HOMOGENEOUS PROTEIN

Under the proper conditions, proteins can be successfully stored for long periods of time. They are best stored as a suspension in concentrated $(NH_4)_2SO_4$ or PEG at 4°C. Concentrated aqueous solutions, 10 mg/mL or higher, with or without 25–50% glycerol, may also be stored. Some investigators filter sterilize the protein solution to reduce microbial degradation (166). Serine transhydroxymethylase is stable for an indefinite period when stored bound to a dye-ligand affinity column of Cibacron blue F3GA at 4°C (185). In an analogous manner, some proteins are stabilized by immobilizing them (186, 187). Enzymes requiring reduced sulfhydryl groups to maintain catalytic activity should be stored with 1-m*M* EDTA and a reduced sulfhydryl reagent over nitrogen or argon. It may be necessary to replenish the sulfhydryl reagent on a regular basis. Many enzymes store well frozen. Small aliquots, 1 mL or less, in 25–50% glycerol can be quick frozen in liquid N_2 and stored in liquid N_2 or in a −80°C freezer. Even cold sensitive enzymes keep well when stored in liquid N_2. Lyophilized proteins, if maintained dry, are stable for extended periods of times. However, many enzymes will not survive the lyophilization process and considerable trial and error, particularly with the buffer conditions, may be required to reconstitute a lyophilized enzyme without significant losses.

ENZYME PURIFICATION

Enzyme ________ Starting Material ________ Weight ________

Date Begun ________ Definition of Activity Unit ________

Protein Method ________ Assay Procedure ________

Procedure	Fraction No.	Vol (mL)	Units/ mL	Total units	Protein (mg/mL)	Specific Activity	Yield (%)	Fold Purifi- cation	Remarks

Fig. 3.9. Example of an enzyme purification summary table.

PROBLEM AREAS

Dilute Protein Solutions

Dilute protein solutions (50 μg/mL or less) should not be stored. Proteins are less stable at low concentrations, multisubunit proteins and cofactors dissociate, and adsorption to surfaces becomes significant. Either the purification step will have to be altered so that the protein is obtained at a higher concentration, or the protein solution that is obtained will have to be concentrated (188) (see Inset 3.2).

If the concentration of the protein solution is above 10 μg/mL, concentrate it by dialysis against 3.8 *M* $(NH_4)_2SO_4$ (Item 1, Inset 3.2). Dialysis against 20% PEG-20,000 may be preferred since PEG-20,000 does not enter the dialysis bag. For protein solutions below 10 μg/mL, remove buffer by adsorption through a dialysis bag using dry powdered PEG-20,000, cellulose, or Sephadex G-100 (Item 3). Surrounding a dialysis bag with dry powdered PEG-20,000, cellulose, or Sephadex G-100, will remove water relatively rapidly. This procedure works well for volumes of 100 mL or less; larger volumes should be divided. The buffer saturated adsorbant can be recovered for later use.

Dried polyacrylate in the form of sticks can be used to concentrate proteins by adding them directly to the protein solution (189). These disposable concentrating sticks absorb 170 mL of H_2O per gram and other low molecular weight substances such as glucose and inorganic salts. Three hundred milligrams of the dried gel concentrates a 50-mL sample to 5mL in 4–5 hr at either 4 or 25°C with good protein recovery. The polyacrylate is formed in 0.6 × 5-cm glass tubes by polymerizing a mixture containing 7% acrylic acid, 0.46% bisacrylamide, 5.3% KOH, 0.02% *N,N,N′,N′*-tetramethylethylenediamine (Temed), and 0.3% ammonium persulfate. After polymerization, the swollen gels are removed from the glass tubes, washed extensively with distilled water to remove persulfate and ultraviolet adsorbing material, and dried in an oven at 50–60°C.

Adsorbing protein from a dilute solution onto a small bed of a chromatographic matrix and subsequently eluting it with a concentrated eluent is a very useful and convenient way to concentrate them (188, 190, 191). If the protein is adsorbed to DEAE cellulose, carboxymethyl dextran is an effective eluting agent because it has a high affinity for this matrix (190). Pressure filtration or ultrafiltration (Item 5) requires special equipment (3); if the equipment is available, it is effective for concentrating

INSET 3.2.

Techniques for Concentrating Protein Solutions

1. Concentrate by dialysis against 3.8 *M* $(NH_4)_2SO_4$ or 25% PEG-20,000 (188).
2. Concentrate large molecular weight proteins by sedimentation with a preparative ultracentrifuge.
3. Place dilute protein solution in dialysis bag and surround with dry powdered PEG-20,000, dry cellulose, or dry Sephadex G-100 (188).
4. Adsorb H_2O directly from the protein solution with dried polyacrylate (189).
5. Adsorb protein to a small bed of a chromatographic matrix and then elute with concentrated eluent (188, 190, 191).
6. Concentrate by pressure filtration through a low molecular weight cutoff membrane filter. Cone shaped low molecular weight cutoff filters are available for use in a centrifugal driven filtration. Pressure filtration through a dialysis bag is also useful (188). Immersible rigid ultrafilters are also available for vacuum filtration and concentration.
7. Concentrate by electrophoretic methods (193–195).
8. Concentrate by lyophilization (188).

both large and small volumes. The equipment required for vaccum filtration through a dialysis bag is easily assembled in the laboratory (188, 192). A variety of electrophoretic systems can also be used to concentrate protein solutions, including systems to concentrate volumes up to 2 mL (193), up to 10 mL (194), or up to 100 mL (195). Finally, dilute protein solutions can be concentrated by lyophilization. As noted earlier, however, the recovery of proteins is generally poor.

Dissociating Cofactors or Subunits

See Chapter 1 for a discussion of this problem and its solution.

Adsorption of Protein to Glass or Plastic

Plastic and glass surfaces adsorb proteins. The lower the protein concentration, the more significant the loss. As discussed in Chapter 1, there are several ways to reduce losses of protein by adsorption to different surfaces: (a) keep the protein solution as concentrated as possible so that the contact surface area of the container or chromatographic matrix is small; (b) the chromatographic support bed should be as small as possible commensurate with good chromatographic procedure; (c) Triton X-100 (0.2 m*M* final concentration or 25–50% glycerol) will often prevent loss of protein by adsorption to surfaces of containers or chromatographic matrices. See Chapter 1 for a more extensive discussion of this problem.

REFERENCES

1. A. L. Demain, *Methods Enzymol.*, **22,** 86–95. (1971).
2. A. Jaworowski, H. D. Campbell, M. I. Poulis, and I. G. Young, *Biochemistry,* **20,** 2041–2047 (1981).
3. T. G. Cooper, *Tools of Biochemistry*, Wiley, New York, 1977, pp. 355–405.
4. W. H. Evans, Preparation and characterization of mammalian plasma membranes, in T. S. Work and E. Work, Eds., *Laboratory Techniques in Biochemistry and Molecular Biology*, Vol. 7, Elsevier–North Holland, Amsterdam, 1979, pp. 11–44.
5. P. J. Hetherington, M. Follows, P. Dunnill, and M. D. Lilly, *Trans. Inst. Chem. Eng.*, **49,** 142–148 (1971).
6. V. Kemerer, C. C. Griffin, and L. Brand, *Methods Enzymol.*, **42,** 91–98 (1975).
7. D. J. South and R. E. Reeves, *Methods Enzymol.*, **42,** 187–191 (1975).
8. O. Warburg and W. Christian, *Biochem. Z.,* **310,** 384–421 (1941).
9. B. L. Taylor, M. F. Utter, *Anal. Biochem.,* **62,** 588–591 (1974).

10. J. W. Thorner, *Methods Enzymol.*, **42,** 148–156 (1975).
11. M. Follows, P. J. Hetherington, P. Dunnill, and M. D. Lilly, *Biotechnol. Bioeng.*, **13,** 549–560 (1971).
12. H. Durchschlag, G. Biedermann, and H. Egger, Eur. *J. Biochem.*, **114,** 255–262 (1981).
13. S-L. Yun, A. E. Aust, and C. H. Suelter, *J. Biol. Chem.*, **251,** 124–128 (1976).
14. J. K. Dethmers, S. Ferguson-Miller, and E. Margoliash, *J. Biol. Chem.*, **254,** 11973–11981 (1979).
15. B. V. Hofsten and A. Tjeder, *Biotechol. Bioeng.*, **3,** 175–180 (1961).
16. T. Achstetter, C. Ehmann, and D. H. Wolf, *Arch. Biochem. Biophys.*, **207,** 445–454 (1981).
17. G. T. James, *Anal. Biochem.*, **86,** 574–579 (1978).
18. V. Sekar and J. H. Hageman, *Biochem. Biophys. Res. Commun.*, **89,** 474–478 (1979).
19. P. H. Morgan, K. A. Walsh, and H. Neurath, *FEBS Lett.*, **41,** 108–110 (1974).
20. P. E. Braid and M. Nix, *Can. J. Biochem.*, **47,** 1–6 (1969).
21. A. P. Volkova and N. R. Grin, *Tr. Tseuntr. Nauchno.-Issled. Dezinfekp, Inst.*, **16–21** (1967); *Chem. Abst.* **70:**10,606m.
22. M. D. Reuber, *Clin. Toxicol.*, **18,** 47–84 (1981).
23. I. Trautschold, E. Werle, and G. Zickgraf-Rudel, *Biochem. Pharm.*, **16,** 59–72 (1967).
24. C. G. Knight and A. J. Barrett, *Biochem. J.*, **155,** 117–125 (1976).
25. A. J. Barrett, Introduction to the history and classification of tissue proteinases, in A. J. Barrett, Ed., *Proteinases in Mammalian Cells and Tissues*, Elsevier–North Holland, Amsterdam, 1977, pp. 1–55.
26. H. Umezawa, *Methods Enzymol.*, **45,** 678–695 (1976).
27. Y. Birk, *Methods Enzymol.*, **45,** 695–697 (1976).
28. R. Laura, D. J. Robison, and D. H. Bing, *Biochemistry,* **19,** 4859–4864 (1980).
29. R. B. Westkaemper and R. H. Abeles, *Biochemistry,* **22,** 3256–3264 (1983).
30. A. Aust, S-L. Yun, and C. H. Suelter, *Methods Enzymol.*, **42,** 176–182 (1975).
31. M. Laskowski, Jr., and R. W. Sealock, *The Enzymes*, 3rd ed., **3,** 375–473 (1971).
32. N. B. Beaty and M. D. Lane, *J. Biol. Chem.*, **257,** 924–929 (1982).
33. A. H. Ramel, Y. M. Rustum, J. G. Jones, and E. A. Barnard, *Biochemistry,* **10,** 3499–3508 (1971).
34. R. Sasaki, E. Sugimoto, and H. Chiba, *Arch. Biochem. Biophys.*, **115,** 53–61 (1966).
35. E. Juni and G. A. Heym, *Arch. Biochem. Biophys.*, **127,** 89–100 (1968).
36. A. E. Aust and C. H. Suelter, *J. Biol. Chem.*, **253,** 7508–7512 (1978).
37. A. Yoshida, *Methods Enzymol.*, **42,** 144–148 (1975).
38. S-L. Yun and C. H. Suelter, *J. Biol. Chem.*, **253,** 404–408 (1978).
39. K-G. Blume and E. Beutler, *Methods Enzymol.*, **42,** 47–53 (1975).
40. J. K. Petell, M. J. Sardo, and H. G. Lebherz, *Prep. Biochem.*, **11,** 69–90 (1981).
41. E. Gohda and H. C. Pitot, *J. Biol. Chem.*, **256,** 2567–2572 (1981).
42. E. H. Ulm, B. M. Pogell, M. M. deMaine, C. B. Libby, and S. J. Benkovic, *Methods Enzymol.*, **42,** 369–374 (1975).
43. D. G. Hardie and P. Cohen, *FEBS Lett.*, **103,** 333–337 (1979).
44. W. D. Loomis and J. Battaile, *Phytochemistry,* **5,** 423–438 (1966).
45. D. D. Randall and N. E. Tolbert, *Methods Enzymol.*, **42,** 405–409 (1975).

46. M. J. A. Tanner, *Curr. Top. Membr. Transp.*, **12,** 1–51 (1979).
47. J. W. Wilson, *Trends Biochem. Sci.*, **3,** 124–125 (1978).
48. A. Helenius and K. Simons, *Proc. Nat. Acad. Sci. USA,* **74,** 529–532 (1977).
49. C. Bordier, *J. Biol. Chem.*, **256,** 1604–1607 (1981).
50. A. Helenius and K. Simons, *Biochim. Biophys. Acta,* **415,** 29–79 (1975).
51. A. Helenius, D. R. McCaslin, E. Fries, and C. Tanford, *Methods Enzymol.*, **56,** 734–749 (1979).
52. C. Tanford and J. A. Reynolds, *Biochim. Biophys. Acta,* **457,** 133–170 (1976).
53. S. Hjerten, H. Pan, and K. Yao, *Protides Biol. Fluids., Proc. Colloq.*, **29,** 15–25 (1982).
54. R. L. Juliano, *Curr. Top. Membr. Transp.*, **11,** 107–144 (1978).
55. H-G. Bock and S. Fleischer, *J. Biol. Chem.*, **250,** 5774–5781 (1975).
56. J-P. Dufour and A. Goffeau, *J. Biol. Chem.*, **255,** 10,591–10,598 (1980).
57. E. Racker, B. Violand, S. O'Neal, M. Alfonzo, and J. Telford, *Arch. Biochem. Biophys.*, **198,** 470–477 (1979).
58. R. K. Scheule and B. J. Gaffney, *Anal. Biochem.*, **117,** 61–66 (1981).
59. S. Clarke, *J. Biol. Chem.*, **250,** 5459–5469 (1975).
60. P. Rosevear, T. van Aken., J. Baxter, and S. Ferguson-Miller, *Biochemistry,* **19,** 4108–4115 (1980).
61. L. M. Hjelmeland, *Proc. Nat. Acad. Sci. USA,* **77,** 6368–6370 (1980).
62. H. W. Chang and E. Bock, *Anal. Biochem.*, **104,** 112–117 (1980).
63. G. E. Tiller, T. Mueller, M. E. Dockter, and W. G. Struve, *Fed. Proc., Fed. Am. Soc. Exp. Biol.*, **41,** 1395 (1982).
64. A. J. Furth, *Anal. Biochem.*, **109,** 207–215 (1980).
65. W. B. Jakoby, *Methods Enzymol.*, **22,** 248–252 (1971).
66. T. H. Massey and W. C. Deal, Jr., *J. Biol. Chem.*, **248**, 56–62 (1973).
67. W. I. Wood, *Anal. Biochem.*, **73,** 250–257 (1976).
68. D. H. Atha and K. C. Ingham, *J. Biol. Chem.*, **256,** 12,108–12,117 (1981).
69. S. I. Miekka and K. C. Ingham, *Arch. Biochem. Biophys.*, **191,** 525–536 (1978).
70. S. I. Miekka and K. C. Ingham, *Arch. Biochem. Biophys.*, **203,** 630–641 (1980).
71. K. C. Ingham and T. F. Busby, *Chem. Eng. Commun.*, **7,** 315–326 (1980).
72. S-C. B. Yan, D. A. Tuason, V. B. Tuason, and W. H. Frey II, *Anal. Biochem.*, **138,** 137–140 (1984).
73. G. E. C. Sims and T. J. Snape, *Anal. Biochem.*, **107,** 60–63 (1980).
74. R. K. Scopes, Techniques for protein purification, in H. L. Kornberg, J. C. Metcalfe, D. H. Northcote, C. I. Pogson and K. F. Tipton, Eds., *Techniques in Protein and Enzyme Biochemistry*, Part I, B101, Elsevier–North, Holland, Amsterdam, 1978, pp. 1–42.
75. R. K. Scopes, K. Griffiths-Smith, and D. G. Millar, *Anal. Biochem.*, **118,** 284–285 (1981).
76. Y. Kato, K. Komiya, Y. Sawada, H. Sasaki, and T. Hashimoto, *J. Chromatogr.*, **190,** 305–310 (1980).
77. M. Rutschmann, L. Kuehn, B. Dahlmann, and H. Reinauer, *Anal. Biochem.*, **124,** 134–138 (1982).

78. F. E. Regnier and K. M. Gooding, *Anal. Biochem.*, **103,** 1–25 (1980).
79. L. Fisher, Gel filtration chromatography, in T. S. Work and R. H. Burdon, Eds., *Laboratory Techniques in Biochemistry and Molecular Biology*, 2nd rev. ed., Elsevier–North Holland, Amsterdam, 1980.
80. A. A. Driedger, L. D. Johnson, and A. M. Marko, *Anal. Biochem.*, **18,** 177–179 (1967).
81. N. Maravalhas, *J. Chromatogr.*, **44,** 617 (1969).
82. E. A. Peterson, Cellulosic ion exchangers, in T. S. Work and E. Work, Eds., *Laboratory Techniques in Biochemistry and Molecular Biology*, Vol. 2, Elsevier–North Holland, Amsterdam, 1970, pp. 225–400.
83. K. L. Smiley, Jr., A. J. Berry, and C. H. Suelter, *J. Biol. Chem.*, **242,** 2502–2506 (1967).
84. I. Takagahara, Y. Suzuki, T. Fujita, J. Yamauti, K. Fujii, J. Yamashita, and T. Horio, *J. Biochem. (Tokyo)*, **83,** 585–597 (1978).
85. D. H. Leaback and H. K. Robinson, *Biochem. Biophys. Res. Commun.*, **67,** 248–254 (1975).
86. C. Chapius-Cellier, A. Francina, and P. Arnaud, *Protides Biol. Fluids Proc. Colloq.*, **27,** 743–746 (1980).
87. J. P. Emond and M. Page, *J. Chromatogr.*, **200,** 57–63 (1980).
88. L. A. AE. Sluyterman and O. Elgersma, *J. Chromatogr.*, **150,** 17–30 (1978).
89. G. Wagner and F. E. Regnier, *Anal. Biochem.*, **126,** 37–43 (1982).
90. R. K. Scopes, *Biochem. J.*, **161,** 253–263 (1977).
91. R. K. Scopes, *Biochem. J.*, **161,** 265–277 (1977).
92. J. R. Davies and R. K. Scopes, *Anal. Biochem.*, **114,** 19–27 (1981).
93. R. K. Scopes, *Anal. Biochem.*, **114,** 8–18 (1981).
94. B. H. J. Hofstee, *Biochem. Biophys. Res. Commun.*, **63,** 618–624 (1975).
95. S. Hjerten, *Adv. Chromatogr.*, **19,** 111–123 (1981).
96. S. Shaltiel, *Methods Enzymol.*, **34,** 126–140 (1974).
97. J. Kohn, R. Lenger, and M. Wilchek, *Appl. Biochem. Biotechnol.*, **8,** 227–235 (1983).
98. S. Hjertén, *Methods Biochem. Anal.*, **27,** 89–108 (1981).
99. K. Nilsson, O. Norrlow, and K. Mosback, *Acta Chem. Scand. Ser. B.*, **35,** 19–27 (1981).
100. K. Nilsson and K. Mosback, *Biochem. Biophys. Res. Commun.*, **102,** 449–457 (1981).
101. M. Einarsson, B. Forsberg, O. Larm, M. E. Requelme, and E. Scholander, *J. Chromatogr.*, **215,** 45–53 (1981).
102. T. Miron and M. Wilchek, *J. Chromatogr.*, **215,** 55–63 (1981).
103. B. H. J. Hofstee, *Biochem. Biophys. Res. Commun.*, **50,** 751–757 (1973).
104. M. Wilchek and T. Miron, *Biochem. Biophys. Res. Commun.*, **72,** 108–113 (1976).
105. G. Halperin, M. Breitenbach, M. Tauber-Finkelstein, and S. Shaltiel, *J. Chromatogr.*, **215,** 211–228 (1981).
106. M. Mevarech, W. Leicht, and M. M. Werber, *Biochemistry*, **15,** 2383–2387 (1976).
107. R. A. Rimerman and G. W. Hatfield, *Science*, **182,** 1268–1270 (1973).
108. W. Leicht and S. Pundak, *Anal. Biochem.*, **114,** 186–192 (1981).
109. K. Vogel, P. Bentley, K-L. Platt, and F. Oesch, *J. Biol. Chem.*, **255,** 9621–9625 (1980).
110. H. H. Varner, Jr., and R. L. Miller, *Prep. Biochem.*, **10,** 1–9 (1980).

111. T. Fujita, Y. Suzuki, J. Yamauti, I. Takagahara, K. Fujii, J. Yamashita, and T. Horio, *J. Biochem.* (*Tokyo*), **87,** 89–100 (1980).
112. S. Shaltiel and Z. Er-El, *Proc. Nat. Acad. Sci. USA,* **70,** 778–781 (1973).
113. Y. Horiuti and S. Imamura, *J. Biochem.* (*Tokyo*), **81,** 1639–1649 (1977).
114. I. Sasaki, H. Gotoh, R. Yamamoto, H. Hasegawa, J. Yamashita, and T. Horio, *J. Biochem.* (*Tokyo*), **86,** 1537–1548 (1979).
115. I Ohkubo, T. Ishibashi, N. Taniguchi, and A. Makita, *Eur. J. Biochem.,* **112,** 111–118 (1980).
116. S. D. Carson and W. H. Konigsberg, *Anal. Biochem.,* **116,** 398–401 (1981).
117. R. Shiman, D. W. Gray, and A. Pater, *J. Biol. Chem.,* **254,** 11,300–11,306 (1979).
118. J. B. Robinson, Jr., J. M. Strottmann, and E. Stellwagen, *Proc. Nat. Acad. Sci. USA,* **78,** 2287–2291 (1981).
119. C. R. Lowe, D. A. P. Small, and A. Atkinson, *Int. J. Biochem.,* **13,** 33–40 (1981).
120. E. Stellwagen, *Acc. Chem. Res.,* **10,** 92–98 (1977).
121. D. A. P. Small, T. Atkinson, and C. R. Lowe, *J. Chromatogr.,* **216,** 175–190 (1981).
122. E. Stellwagen, *Colloques de l'Inserm* (*INSERM*), **86,** 345–356 (1979).
123. P. D. G. Dean and D. H. Watson, *J. Chromatogr.,* **165,** 301–319 (1979).
124. R. F. Tucker, J. Babul, and E. Stellwagen, *J. Biol. Chem.,* **256,** 10,993–10,998 (1981).
125. M. Sugiura, *Anal. Biochem.,* **108,** 227–229 (1980).
126. C. J. Bruton and T. Atkinson, *Nucleic Acid Res.,* **7,** 1579–1591 (1979).
127. J. B. Robinson, Jr., J. M. Strottmann, D. G. Wick, and E. Stellwagen, *Proc. Nat. Acad. Sci. USA,* **77,** 5847–5851 (1980).
128. G. Jones and M. Dole, *J. Am. Chem. Soc.,* **51,** 2950–2964 (1929).
129. R. K. Scopes, *Anal. Biochem.,* **136,** 525–529 (1984).
130. C. R. Lowe, An introduction to affinity chromatography, in T. S. Work and E. Work, Eds., *Laboratory Techniques in Biochemistry and Molecular Biology,* Vol. 7, Elsevier–North Holland, Amsterdam, 1979, pp. 269–518.
131. W. H. Scouten, *Affinity Chromatography: Bioselective Adsorption on Inert Matrices,* Wiley, New York, 1981.
132. R. Harrison, M. Rangarajan, A. Dell, D. Mercer, R. H. Pain, et al., Structural investigations of peptides and proteins, in R. C. Sheppard, Ed., *Amino-Acids, Peptides, and Proteins,* Vol. 10, Burlington House, London, 1979 pp. 33–45.
133. D. R. Absolom. *Sep. Purif. Methods,* **10,** 239–286 (1981).
134. M. R. Sairam., *J. Chromatogr.,* **215,** 143–152 (1981).
135. L. S. Nielsen, J. G. Hansen, L. Skriver, E. L. Wilson, K. Kaltoft, J. Zeuthen, and K. Dano, *Biochemistry,* **21,** 6410–6415 (1982).
136. R. Bartholomew, D. Beidler, and G. David, *Protides Biol. Fluids, Proc. Colloq.,* **30,** 667–670 (1983).
137. C. Schneider, R. A. Newman, D. R. Sutherland, U. Asser, and M. F. Greaves, *J. Biol. Chem.,* **257,** 10,766–10,769 (1982).
138. M. Sugiura, S. Hayakawa, T. Adachi, Y. Ito, K. Hirano, and S. Sawaki, *J. Biochem. Biophys. Methods,* **5,** 243–249 (1981).
139. J. M. MacSween and S. L. Eastwood, *Methods Enzymol.,* **73,** 459–471 (1981).
140. M-T. Haeuptle, M. L. Aubert, J. Djiane, and J-P. Kraehenbuhl, *J. Biol. Chem.,* **258,** 305–314 (1983).

141. R. F. Murphy, A. Imam, A. E. Hughes, M. J. McGucken, K. D. Buchanan, J. M. Conlon, and D. T. Elmore, *Biochim. Biophys. Acta,* **420,** 87–96 (1976).

142. F. E. Regnier, *Anal. Biochem.,* **126,** 1–7 (1982).

143. T. W. Hearn, F. E. Regnier, and C. T. Wehr, *Am. Lab.,* **14,** 18–39 (1982).

144. C. R. Lowe, M. Glad, P.-O. Larsson, S. Ohlson, D. A. P. Small, T. Atkinson, and K. Mosback, *J. Chromatogr.,* **215,** 303–316 (1981).

145. P.-O. Larsson, T. Griffin, and K. Mosback, *Colloques de l'Inserm* (*INSERM*), **86,** 91–98 (1979).

146. Y. Kato, T. Kitamura, and T. Hashimoto, *J. Chromatogr.* **266,** 49–54 (1983).

147. L. Soderberg et al., *Protides Biol. Fluids, Proc. Colloq.,* **30,** 629–634 (1983).

148. J. Bergstrom. L. Soderberg, L. Wahlstrom, R.-M. Muller, A. Domicelj, G. Hagstrom, R. Stalberg, I. Kallman, and K.-A. Hansson, *Protides Biol. Fluids, Proc. Colloq.,* **30,** 641–646 (1983).

149. R. R. Bott, M. A. Navia, and J. L. Smith, *J. Biol. Chem.,* **257,** 9883–9886 (1982).

150. E. M. Southern, *Anal. Biochem.,* **100,** 304–318 (1979).

151. S. A. Saeed and T. R. C. Boyde, *Prep. Biochem.,* **10,** 445–462 (1980).

152. S. Otsuka and I. Listowsky, *Anal. Biochem.,* **102,** 419–422 (1980).

153. C. Bernabeu, F. Sanchez-Madrid, and R. Amils, *Eur. J. Biochem.,* **109,** 285–290 (1980).

154. D. A. Hager and R. R. Burgess, *Anal. Biochem.,* **109,** 76–86 (1980).

155. J. K. W. Mardian and I. Isenberg, *Anal. Biochem.,* **91,** 1–12 (1978).

156. E. Mendez, *Anal. Biochem.,* **126,** 403–408 (1982).

157. K. P. Brooks and E. G. Sander, *Anal. Biochem.,* **107,** 182–186 (1980).

158. A. Chrambach and D. Rodbard, *Science,* **172,** 440–451 (1971).

159. J. A. Braatz and K. R. McIntire, *Prep. Biochem.,* **7,** 495–509 (1977).

160. M. Otto and M. Snejdarkova, *Anal. Biochem.,* **111,** 111–114 (1981).

161. P. G. Righetti and E. Gianazza, *J. Chromatogr.,* **184,** 415–456 (1980).

162. D. A. Daglish, M. T. W. Hearn, A. J. Paterson, R. L. Prestidge, and P. G. Stanton, *Prep. Biochem.,* **11,** 201–216 (1981).

163. J. E. Fuhr, S. Pavlovic, and R. Bovino, *Biochem. Biophys. Res. Commun.,* **98,** 930–935 (1981).

164. C. Chapuis-Cellier and P. Arnaud, *Anal. Biochem.,* **113,** 325–331 (1981).

165. B. M. Harpel and F. Kueppers, *Anal. Biochem.,* **104,** 173–174 (1980).

166. D. L. Pierson and J. M. Brien, *J. Biol. Chem.,* **255,** 7891–7895 (1980).

167. A. D. Friesen, J. C. Jamieson, and F. E. Ashton, *Anal. Biochem.,* **41,** 149–157 (1971).

168. W. F. Goldman and J. N. Baptist, *J. Chromatogr.,* **179,** 330–332 (1979).

169. B. D. Hames and D. Rickwood, Eds., *Gel Electrophoresis of Proteins: A Practical Approach, IRL Press,* Oxford, (1981).

170. S. D. Flanagan and S. H. Barondes, *J. Biol. Chem.,* **250,** 1484–1489 (1975).

171. P.-A. Albertsson, *J. Chromatogr.,* **159,** 111–122 (1978).

172. P. Hubert and E. Dellacherie, *J. Chromatogr.,* **184,** 325–333 (1980).

173. G. Kopperschlager and G. Johansson, *Anal. Biochem.* **124,** 117–124 (1982).

174. P. O. Hedman and J-G. Gustafsson, *Anal. Biochem.,* **138,** 411–415 (1984).

175. M. Spencer and M. Grynpas, *J. Chromatogr.,* **166,** 423–434 (1978).

176. D. A. Hillson, *J. Biochem. Biophys. Methods,* **4,** 101–111 (1981).
177. H. Hansson and L. Kagedal, *J. Chromatogr.,* **215,** 333–339 (1981).
178. W. B. Jakoby, *Methods Enzymol.,* **22,** 248–252 (1971).
179. A. McPherson, Jr., *J. Biol. Chem.,* **251,** 6300–6303 (1976).
180. C. Thaller, L. H. Weaver, G. Eichele, E. Wilson, and R. Karlsson, *J. Mol. Biol.,* **147,** 465–470 (1981).
181. T. Alber, F. C. Hartman, R. M. Johnson, G. A. Petsko and D. Tsernglou, *J. Biol. Chem.,* **256,** 1356–1361 (1981).
182. D. R. Davies and D. M. Segal, *Methods Enzymol.,* **22,** 266–269 (1971).
183. A. McPherson, *Preparation and Analysis of Protein Crystals,* Wiley, New York, 1982.
184. G. Sofer and V. J. Britton, *BioTechniques,* **1,** 198–203(1983).
185. J. C. Braman, M. J. Black, J. H. Mangum, *Prep. Biochem.,* **11,** 23–32 (1981).
186. K. Mosbach, *FEBS Lett.,* **62** (Supplement), E80–E95 (1976).
187. J. Schnapp and Y. Shalitin, *Biochem. Biophys. Res. Commun.,* **70,** 8–14 (1976).
188. W. F. Blatt, *Methods Enzymol.,* **22,** 39–49 (1971).
189. H. G. Vartak, M. V. Rele, M. Rao, and V. V. Deshpande, *Anal. Biochem.,* **133,** 260–263 (1983).
190. A. R. Torres and E. A. Peterson, *Anal. Biochem.,* **98,** 353–357 (1979).
191. E. Schmincke-Ott and H. Bisswanger, *Prep. Biochem.,* **10,** 69–75 (1980).
192. W. J. Hubbard, *Prep. Biochem.,* **7,** 313–319 (1977).
193. D. G. Rhodes and D. A. Yphantis, *Anal. Biochem.,* **116,** 379–382 (1981).
194. W. B. Allington, A. L. Cordry, G. A. McCullough, D. E. Mitchell, and J. W. Nelson, *Anal. Biochem.,* **85,** 188–196 (1978).
195. I. Posner, *Anal. Biochem.,* **70,** 187–194 (1976).

4

Physical and Chemical Properties of a Protein

Studying the physical and chemical properties of a protein is a rewarding experience. It can be likened to a tour of the streets of Rome; a study of the buildings and peoples of this great city provides an appreciation of its functions and how they have evolved over the years. Likewise, a study of the physical and chemical properties of a protein molecule provides a greater appreciation of its function. In this chapter, I shall briefly describe some of the tools available for such a study. Some of the important functions of a protein are discussed in the next chapter.

CRITERIA FOR HOMOGENEITY

One of the first characteristics of a newly isolated protein preparation of interest to most enzymologists is its homogeneity. Is the preparation contaminated with other proteins? How pure is it? In the past, enzymologists expended considerable time and effort attempting to demonstrate the homogeneity or purity of a new protein preparation. No longer is this necessary; polyacrylamide gel electrophoresis has simplified the process. By subjecting 50–100 μg of protein to polyacrylamide gel electrophoresis and staining the protein in the gel with a suitable stain, it is possible to detect less than 1% contamination.

To be sure that the preparation is homogeneous, examine the electrophoretograms of the native preparation at 2 or 3 pH values and at several acrylamide concentrations. A homogeneous preparation should migrate as a single band under all conditions. Using different conditions increases the possibility of detecting contaminating proteins, since it is unlikely that they will have the same electrophoretic mobility under all conditions as the desired protein. Electrophoretograms of sodium dodecyl sulfate dissociated proteins will give more than one band if the protein is composed of different sized subunits. Sodium dodecyl sulfate polyacrylamide gel electrophoresis determines homogeneity of subunits not native proteins.

It is also important to correlate enzyme activity with the protein band on the polyacrylamide gel because it is possible that the enzyme activity of interest may be associated with a minor protein component in the preparation. After running duplicate gels, stain one for protein and the other for enzyme activity, or slice the duplicate gel into 1-mm slices and extract the protein by macerating each slice in a small volume of buffer. The extract can then be assayed for enzyme activity. Some investigators slice a gel longitudinally and stain one half with a protein stain and the other half with an enzyme activity stain. It is also possible to renature the Coomassie Brilliant Blue stained proteins in situ and then stain for enzymic activity (1).

It is now known from polyacrylamide gel electrophoretic data that it is difficult to isolate an enzyme completely free of contaminating protein. For most experiments, however, a 5% contamination with other proteins is acceptable. On the other hand, some protein preparations, such as enzymes or hormones that are to be administered intravenously, must have much less contamination. Human insulin, which is synthesized with a microbial system using recombinant deoxyribonucleic acid (DNA) techniques, for example, must be antigenically pure. In other cases, it is necessary to determine whether the enzyme preparation is contaminated with other enzyme activities. How the protein is used then dictates acceptable protein contamination.

ABSORBANCY INDEX

Protein concentrations are routinely determined by a variety of empirical methods as discussed in Chapter 2. When working with homogeneous proteins, however, empirical methods often give incorrect concentrations, and, therefore, I recommend using absorbance measurements and the absorbancy index ϵ to determine their concentration. The absorbancy index is ordinarily defined as the absorbance of a solution containing 1 mg/mL protein. Most absorbance measurements are made at 280 nm.

The only precise method of determining the absorbancy index of a protein is to determine the dry weight of the protein in a sample of known absorbance (2). For proteins soluble in distilled water, dialyze extensively against distilled water, centrifuge to remove fine particulate material, and measure the absorbance. An aliquot of the protein solution containing 5–10 mg protein and an identical aliquot of the dialyzate are then placed in preweighed drying flasks and dried at 105°C for at least 24 hr, allowed to cool in a desiccator over P_2O_5, and weighed with an analytical balance, preferably one sensitive at the microgram level. Because dried protein absorbs water, the weighing must be done quickly. Repeat this process of drying and cooling until constant weights are obtained. Of course, good weighing techniques must be employed.

If the protein is not soluble in distilled water, a volatile or nonvolatile salt or buffer must be added to the sample before dialysis. A nonvolatile salt or buffer is recommended because sublimation of a volatile salt often carries protein with it. The final weight of the dried protein must be corrected for the salt in the dialyzate.

The problems encountered when determining the dry weight of protein in a sample are reviewed in Reference (2). Generally, use the smallest vial commensurate with the size of the sample to be dried. Placing small volume samples—200 μL or less—on glass fiber discs enhances their drying rate. Because of the excessive amounts of material required to determine the amount of protein in a sample by dry weight measurements, some investigators estimate its concentration by several of the empirical methods given in Table 2.1 and then calculate an average extinction coefficient (3).

Because the dry weight procedure requires up to 10 mg of protein and a minimum of 24 hr, Kickhofen and Warth (4) devised a shorter procedure using an electrobalance, which requires 0.25 to 2 mg of protein and 40 min for completion. In this procedure, a small aliquot, 50–200 μL, of both the protein solution and the dialyzate are pipetted onto glass fiber discs and dried under vacuum with a specially designed balance. The samples are allowed to dry to constant weight as indicated by a recorder interfaced to the electrobalance.

ACTIVE SITE CONCENTRATION

The amount of enzyme in a crude extract or in a purified preparation is usually determined by a rate assay as described in Chapter 2. This type of assay works well when devising a purification scheme and for most routine measurements. Rate assays are fraught with many uncertainties, however, which limit their accuracy and reproducibility from laboratory to laboratory. Temperature, ionic strength, substrate concentration, cofactors, inhibitors, activators, pH, and so on, must all be controlled; in short, rate assays lack a standard. To overcome these difficulties, Bender et al. (5) suggest determining the concentration of enzyme active sites. The amount of enzyme in solution is then expressed in terms of a normality, that is, active site equivalents per liter.

In principle, three different kinds of procedures, based on different strategies, are commonly used to determine the normality of an enzyme solution. They are based on: (a) reacting the enzyme with a radioactive competitive inhibitor that forms a covalent bond with it; (b) titrating the enzyme with a very strong binding cofactor, inhibitor, or transition state analog by following changes in absorbance or fluorescence of the ligand; and (c) adding a substrate that reacts with the enzyme in two distinct kinetic phases—the extent of the first phase gives a measure of the active site concentration. An example of each procedure will be discussed.

Santi et al. (6) describe an example of the first approach in which a radioactive competitive inhibitor forms a covalent bond with an enzyme. 5-Fluoro-2-deoxyuridylate, in the presence of 5,10-methylenetetrahydrofolic acid, forms a covalent bond with thymidylate synthetase. The binding is rapid, stoichiometric, and exhibits a high or absolute specificity. The enzyme is added to a reaction mixture containing buffer, a sulfhydryl reducing agent, the necessary substrates and the radioactive 5-fluoro-2-deoxyuridylate. An aliquot of this solution is then applied to a nitrocellulose filter that retains the protein and allows the unreacted reagents and other reaction components to pass through. After washing the excess unreacted radiolabeled material from the filter, the filter is counted in a scintillation spectrometer. The sensitivity of the assay appears to be limited by the specific activity of the labeled inhibitor; as little as 3×10^{-14} mol of active site of thymidylate synthetase have been quantitated (6). Transition state analogs (7) are not covalently bound, but they have a high affinity for the enzyme so that the analog protein complex can also be quantified by retention on a nitrocellulose filter.

The second approach is commonly used to determine the normality of horse liver alcohol dehydrogenase (8, 9). Adding nicotinamide adenine dinucleotide (NAD^+) to a sample of alcohol dehydrogenase in the presence of the inhibitor, pyrazole, results in the formation of a ternary com-

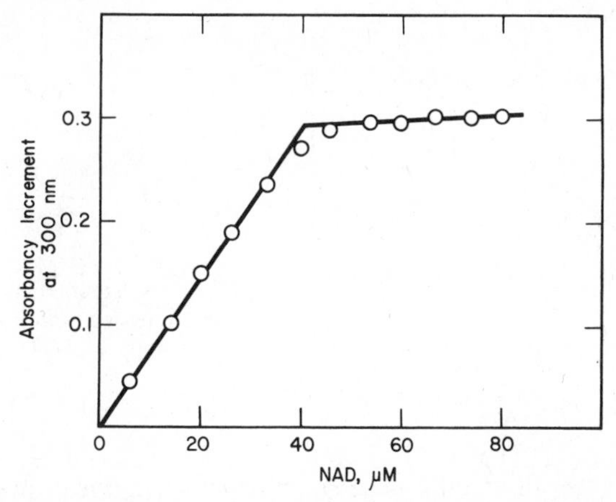

Fig. 4.1. A spectrophotometric titration of 40 μN liver alcohol dehydrogenase with NAD^+ in the presence of 660 μM pyrazole. Adapted with permission from Reference (8).

plex, NAD^+–pyrazole–enzyme, which absorbs at 300 nm, ϵ_{300} = 7.2/mM cm. When this change in absorbance is plotted as a function of the amount of NAD^+ added (Figure 4.1), the saturation point at the break in the curve gives the concentration of the enzyme directly. A method similar to this involves titration of alcohol dehydrogenase with nicotinamide adenine dinucleotide (reduced) (NADH) in the presence of isobutyramide; plotting the change in fluorescence as a function of added NADH gives a plot identical in shape to Figure 4.1 (9).

For the titration procedures used in the second approach to function properly, the dissociation constant for the titrant must be 50 to 100 times smaller than the concentration of the active sites to be titrated. If the K_D is significantly larger than this, a sharp break will not be observed, and the concentration of the active site cannot be accurately discerned. Of course, the concentration of the enzyme must also be large enough to give a significant absorption or fluorescence signal.

The third approach to determining the normality of enzyme active sites is a kinetic approach, which is often used with serine esterases (5). Equation (4.1) describes the overall process for serine esterases:

$$E + S \underset{}{\overset{K_s}{\rightleftharpoons}} ES \underset{k_{-2}}{\overset{k_2}{\rightleftharpoons}} ES_1 \underset{\downarrow P_1}{\overset{k_3}{\longrightarrow}} EP_2 \underset{k_{-4}}{\overset{k_4}{\rightleftharpoons}} E + P_2 \tag{4.1}$$

If the product P_1, formed by the reaction of enzyme E with substrate S, can be measured before appreciable breakdown of EP_2 occurs to regen-

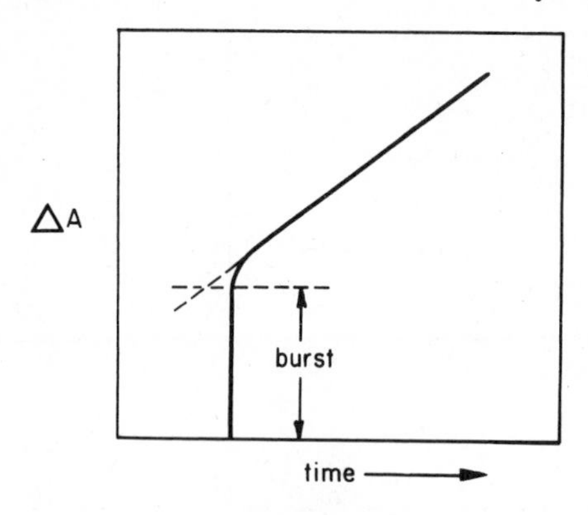

Fig. 4.2. A hypothetical curve showing the rapid release of *p*-nitrophenol after addition of *p*-nitrophenylacetate to chymotrypsin.

erate the enzyme, then a direct measure of the enzyme concentration can be made. For example, the reaction between chymotrypsin and *p*-nitrophenyl acetate proceeds in two distinct kinetic steps, as described by Equation (4.1) and indicated in Figure 4.2. First there is a fast initial liberation of approximately 1 mol of *p*-nitrophenol per mole of enzyme, followed by a slow reaction. The slow second step is due to the hydrolysis of the acyl enzyme, EP_2, which regenerates the catalyst. The extent of the first fast step gives a measure of the active site concentration.

The main difficulty with this assay is the lack of sensitivity; the lowest concentration of *p*-nitrophenol that can be measured with an accuracy of better than 2% is 10^{-6} *M*. To overcome this problem, Livingston et al. (10) synthesized a new active site titrant, shown in Figure 4.3. In the presence of trypsin, one of the guanidinobenzoyl esters of the active site titrant is hydrolyzed to form an acyl enzyme derivative, *p*-guanidinobenzoyl enzyme, and a highly fluorescent derivative of fluorescein. The acyl enzyme derivative analogous to EP_2 in Equation (4.1), that is, *p*-guanidinobenzoyltrypsin, has a half-time for regeneration of 12 hr; this long half-life insures distinct and observable phases. Because the product P_1 is highly fluorescent, this reagent allows determination of concentrations of enzyme active sites as low as 10^{-12} *M*.

Kezdy and Kaiser (11) emphasize several points that should be considered in establishing a valid assay of enzyme active sites: the reactant must have a one-to-one stoichiometry with the active site of the enzyme; the reagent should allow a saturation phenomenon; substrates and competitive inhibitors should inhibit the titration reaction; titrants that react covalently should destroy enzyme catalytic activity; reagents that possess an asymmetric center should show a large difference in reactivity; and

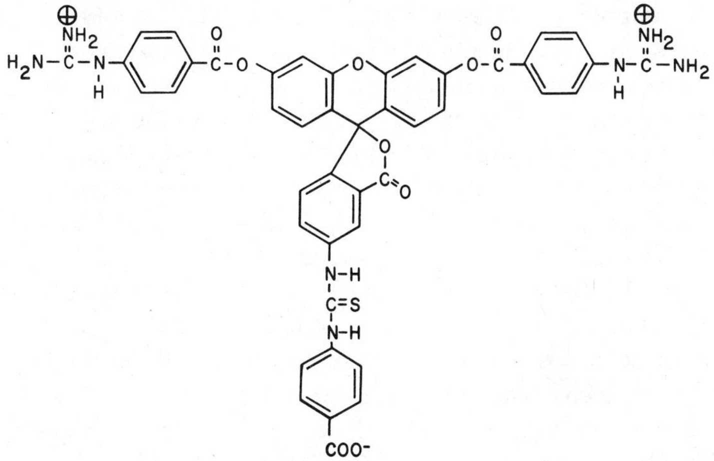

Fig. 4.3. Structure of guanidinobenzoyl ester 3′, 6′-bis(4-guanidinobenzoyloxy)-5-[*N*′-(4-carboxyphenyl) thioureido]spiro[isobenzofuran-1-(3H),9′[9H] xanthen]-3-one (10).

experimental conditions such as 8 *M* urea or pH extremes that alter the three-dimensional conformation of the groups at the active site should prevent the titration.

THE ISOELECTRIC pH

Proteins are ampholytes, meaning each molecule contains both acidic and basic residues. The isoelectric pH or isoelectric point (pI) is the pH at which the number of positively charged groups equals the number of negatively charged groups on the protein surface. At this pH, proteins have zero mobility in an electric field. Often this pH will also dictate the behavior of proteins on an ion exchange chromatographic medium (12).

The pI of a protein is estimated by isoelectric focusing or by electrophoresis at different pH values. Isoelectric focusing, the most popular method for determining the pI, uses a density gradient of sucrose in a small column (13) or a more viscous gel medium, such as polyacrylamide or special agarose on a flat bed (14). This discussion will be limited to isoelectric focusing experiments in polyacrylamide gels.

Some of the critical parameters involved in determining the pI of a protein in a polyacrylamide gel medium are discussed in References (15)–(20). First, use the highest voltage possible. The higher the voltage, the faster the migration rates of both the carrier ampholytes and the proteins. The increased rate of migration reduces both the time that proteins are exposed to harsh experimental conditions and the time necessary to complete an experiment. Higher voltages also increase the resolution of the technique by reducing the size of the focused protein zone.

The composition of the carrier ampholytes and the effectiveness of the equipment in cooling the gel are the main factors involved in limiting the voltage. If the carrier ampholytes have an uneven conductivity over their pH range, localized overheating may result causing water evaporation, gel shrinkage, and heat denaturation of proteins. The equipment, therefore, must be effective in uniformly removing heat from the gel plate.

One experimental difficulty with some electrofocusing experiments is gradient drift, a decrease in pH with time at the cathode and a flattening of the gradient at the anode (18). Gradient drift should not be a problem unless extensive times are required to focus the proteins. The drift can be reduced by using acrylamide that has been recrystallized or passed over a mixed bed ion exchanger to remove contaminating acrylic acid or by substituting amino acids or buffers for either one of the usual strongly acidic anolyte or strongly basic catholyte or both. For the following pH ranges, the indicated anolyte–catholyte may be used: for pH 2.5–5, use 0.1 *M* H_2SO_4–0.2 *M* histidine; for pH 4–6.5, use 0.04 *M* glutamic acid–0.2 *M* histidine; for pH 5–8, use 0.04 *M* glutamic acid–1 *M* NaOH; for pH 5–9, use 0.25 *M* *N*-(2-hydroxyethyl)piperazine-*N′*-2-ethanesulfonic acid (Hepes)–1 *M* NaOH; and for pH 3–10, use 0.04 *M* aspartic acid–1 *M* NaOH (14). When pH gradients are established with mixtures of simple buffers (19) instead of commercial or synthetic ampholytes, use the most acidic buffer as the anolyte and the most basic as the catholyte.

Laas et al. (15) have shown that despite the use of one of the best isoelectric focusing devices available and an excellent ampholyte, the temperature at various points over a gel plate still varies significantly. Furthermore, temperatures at the electrodes often rise at the beginning of an experiment, so consequently, proteins should either not be applied in these regions or else applied after prefocusing the gel. Because the pI of a protein and the pH of the carrier ampholyte vary with temperature, the pH of the focused ampholytes should be determined at the same temperature that the gel was maintained at during the experiment. Finally, apply a minimum amount of protein to the isoelectric focusing gel. Proteins are also ampholytes, so excessive sample will distort the pH gradient (21). Including a nonionic detergent (Brij 35) in the gel matrix will minimize the tendency of some proteins to precipitate at their isoelectric point (22). Although the procedures recommended by Laas et al. (15) for determining an accurate pI value take many of these points into consideration, a convenient way to obtain reasonably accurate pI values is to prepare a standard curve by plotting the positions of several standard proteins on the focused gel versus their pI values. Mixtures of such standard proteins are commercially available.

Isoelectric focusing gels already cast on flexible plastic sheets are commercially available and convenient to use but expensive. To offset the expense, similar 1-mm thick polyacrylamide gels on glass plates can easily be prepared. Treating glass plates first with γ-methacryloxypropyl trimethoxysilane (Silane A-174) makes it possible to covalently bind the gel to the plate surface. The trimethoxysilane portion of Silane A-174 reacts covalently with the glass leaving the methacrylate portion available to participate in the acrylamide polymerization reaction (15). To reduce the restriction on the movement of proteins, particularly large proteins, in the matrix, a low percentage gel is recommended (17).

To further reduce the cost of an electrofocusing experiment, some investigators synthesize their own ampholytes (23, 24) or, as noted above, use a mixture of buffers with different p*K* values (19). Polybuffer may also be substituted for ampholytes (25). An even less expensive but also more time consuming method for determining the pI of a protein does not require ampholytes, instead it uses cellulose acetate electrophoresis at a series of pH values. The pI equals the pH at which electrophoretic mobility equals zero. A list of 400 proteins and their pI values is given in Reference (26).

ABSORPTION OF ULTRAVIOLET LIGHT

Because proteins absorb ultraviolet (UV) light, spectrophotometric measurements are indispensable in their study. Spectrophotometric data are used: (a) to quantify protein (see Chapter 2), (b) to determine the number of aromatic amino acids in a protein, (c) to monitor the titration of ionizable groups, (d) to study the topology of a protein, (e) to monitor protein–protein or protein–ligand interactions, (f) to monitor the denaturation of protein, (g) to monitor protein conformation changes, and (h) to predict elements of its secondary structure (27, 28).

Table 4.1 Wavelength maxima and molar extinction coefficients of protein chromophoric groups

	Solvating conditions					
	Neutral pH		Alkaline pH		Nonpolar[a]	
Chromophore	(nm)	$(M\ \mathrm{cm})^{-1}$	(nm)	$(M\ \mathrm{cm})^{-1}$	(nm)	$(M\ \mathrm{cm})^{-1}$
Peptide bond[b,c]	190	7,000	190	4,300		
Cysteine[c]	190	100	238	5,000		
Imidazole[c]	212	6,000				
Tyrosine[d]	275	1,405	293	2,381	278	1,790
	223	8,260	240	11,340		
Tryptophan[d]	279	5,579	280	5,385	282	6,170
	219	46,700				
Phenylalanine[d]	252	154	253	171	252	158
	258	195	258	210	258	195
	264	151	264	161	264	155
	267	91	268	118	268	96
	206	9,340				

[a] For these measurements, amino acids were dissolved in ethanol.
[b] The parameters for the peptide bond are obtained with poly(L-lysine).
[c] Reference (29).
[d] From Reference (27) by permission.

The major chromophoric groups in a protein are given in Table 4.1. The peptide bond because of its concentration, as in poly(L-lysine), is responsible for most of the absorption at 190 nm. The decrease in the extinction coefficient of the peptide bond of poly(L-lysine) at alkaline pH values, as shown in Figure 4.4, is due to the α helix that forms when this peptide is neutralized. Note that the wavelength of maximum absorption does not change.

Three of the amino acid residues—the imidazolium, the sulfhydryl, and the tyrosyl residues—undergo significant spectral changes when titrated with base. Ionizing the imidazolium residue produces a negative difference peak at 226 nm [$\Delta\epsilon = -285(M\ \mathrm{cm})^{-1}$] (data not shown). The absorption maximum of the sulfydryl residue shifts from 190 nm [$\epsilon = 100(M\ \mathrm{cm})^{-1}$] at neutral pH values to 238 nm [$\epsilon = 5000(M\ \mathrm{cm})^{-1}$] after ionization at alkaline pH values. Ionizing the tyrosyl hydroxyl group causes both the short wavelength and the long wavelength absorption maxima to shift

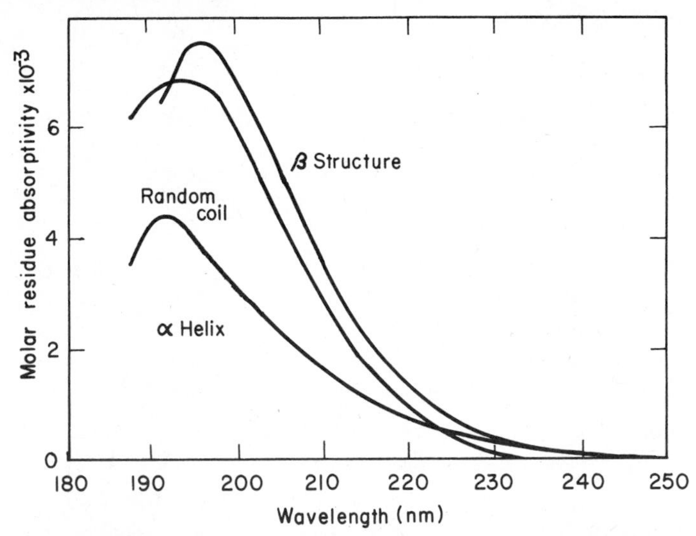

Fig. 4.4 Far ultraviolet absorption spectra of poly(L-lysine) in aqueous solution: random coil, pH 6.0, 25°C; α helix, pH 10.8, 25°C; and β structure, pH 10.8, 52°C. Abstracted by permission of the author. Reference (29).

to longer wavelengths and the extinction coefficients at both wavelengths to increase. Tryptophanyl and phenylalanyl absorption maxima are not significantly affected by increases in pH (Table 4.1).

Tyrosine and tryptophan absorb the majority of light in the 280-nm region. Phenylalanine, because of its low extinction coefficient, and cysteine and histidine, because of their low concentration, do not contribute significantly to the absorbance of most proteins in the UV region. Figure 4.5 shows the absorption spectra of tyrosine, tryptophan, and phenylalanine. (Note that the vibronic transitions of phenylalanine are typical of those of benzene.) As indicated in Table 4.1, none of the transitions of phenylalanine are affected by changes in solvent or pH. In contrast, the positions of the absorption bands of tyrosine and tryptophan are affected by solvent polarity. The 280-nm band of tyrosine occurs at 282 nm in a polar medium (water–glycerol, 1:1) and near 288 nm in nonpolar solvents. The tryptophanyl spectrum consists of two overlapping electronic transitions. Both are affected by solvent as will be discussed more extensively in later sections of this chapter. Consult Reference (27) for an extended analysis of the spectral properties of the three aromatic amino acids.

Measuring changes in the absorption spectrum of a protein by difference spectroscopy is more accurate than those obtained by manually sub-

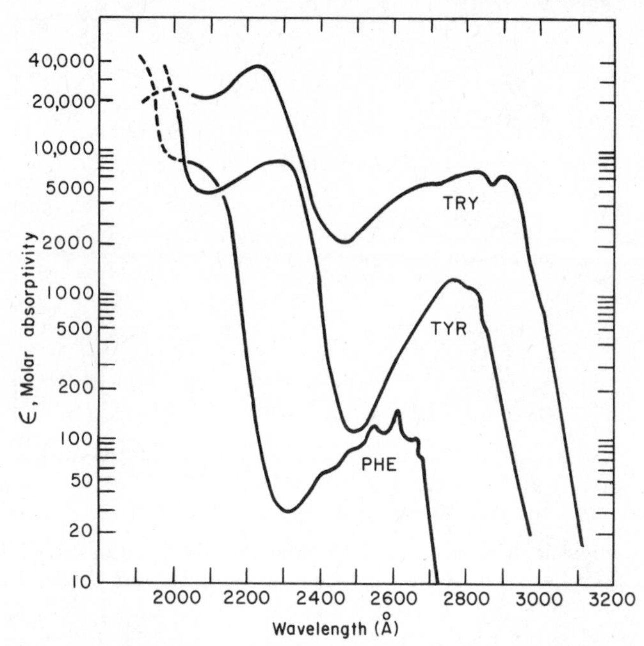

Fig. 4.5. Absorption spectra of the aromatic amino acids tryptophan (Try), tyrosine (Tyr), and phenylalanine (Phe) at pH 6. Abstracted with permission from Reference (30).

tracting one spectral measurement from another. Differences in absorption are measured directly after placing a protein solution in both the reference and sample position of the spectrophotometer. Perturbing the protein in the sample position by changing the pH, temperature, or polarity of the solvent, or by adding specific ligands may alter its absorption spectrum. If the perturbation changes the protein conformation, one or more buried tryptophanyl or tyrosyl residues may be exposed to the bulk solvent causing a change in their absorption spectra. Changes in pH may change the ionization of various functional groups on the protein surface, which in turn may affect the protein's three-dimensional structure. Changing the temperature or adding ligands that interact at specific sites on the protein may also affect its conformation. In addition, the absorption spectrum of the chromophoric groups can be altered directly by ionization or protonation. Adding a nonpolar solvent, such as ethylene glycol, changes the absorption spectrum of surface chromophoric groups by altering the polarity of the solvent in contact with them.

Figure 4.6 presents the difference spectra of several amino acids perturbed by 20% ethylene glycol. Of the five difference spectra presented,

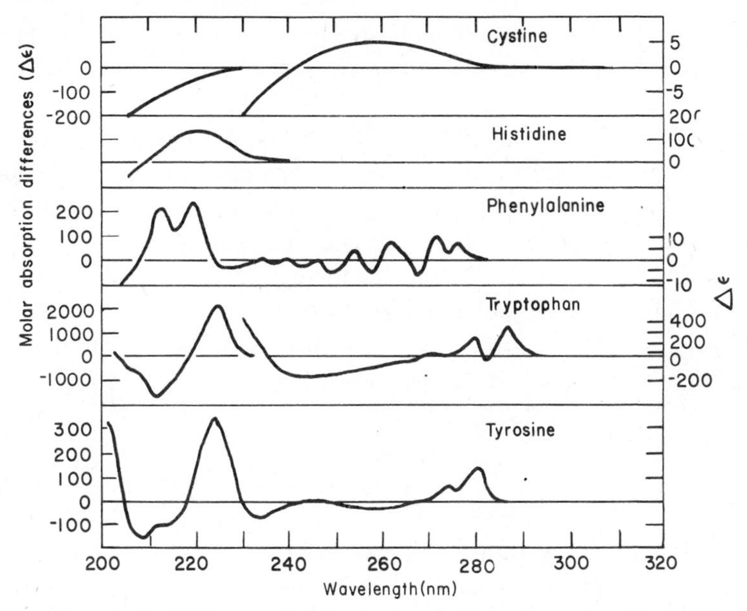

Fig. 4.6. Perturbation difference spectra of some amino acids produced by 20% (v/v) ethylene glycol. The absorption at all difference peaks obeys Beer's law at the lower concentrations employed (2×10^{-3} to 2×10^{-5} *M*). When two scales are shown for ε, the ordinate on the right applies to the right-hand portion of the curve. Abstracted with permission from Reference (31).

those produced by tyrosine and tryptophan give the largest differences and are most useful. Note the major difference peaks for tryptophan are at 224, 282, and 290 nm; for tyrosine, the peaks are at 223, 278, and 283 nm. The wavelength maxima of difference absorbance peaks observed when the tryptophanyl residues buried in the interior of the protein molecule are exposed to solvent during a conformation change, are usually shifted 5 nm to longer wavelengths; peaks or wells at 287 and 295 nm are usually observed in these difference spectra (27).

Light is an electromagnetic wave consisting of an oscillating electric field and a magnetic field, both of which can be represented by mutually perpendicular vectors. A typical light source emits randomly oriented oscillating vectors so that the emitted light is a collection of waves with all possible orientations. By passing light through prisms or devices that transmit in only one vector plane, plane-polarized light is obtained; circularly polarized light is obtained by superimposing two plane polarized light waves that differ in phase by one-quarter wavelength (32). The light wave is either right or left circularly polarized depending on whether the vector rotates clockwise or counter clockwise when viewed by an observer looking at the source.

Either the velocity or amplitude of an oscillating vector can be reduced when the light is allowed to pass through certain substances. Reducing the velocity of the wave propagation is called refraction, denoted by η; reducing the amplitude of the vector is called absorption, denoted by ϵ.

When the substance through which the light passes is optically active, right and left circularly polarized vectors will be affected differently. If an unequal reduction in the velocity of the right and left circularly polarized light is assumed, it can be shown that

$$\alpha = \frac{180\,d}{\lambda}(\eta_L - \eta_R)$$

where α is the optical rotation at a specific wavelength λ, d is the light path in decimeters, and η_L and η_R are the indices of refraction of the left and right circularly polarized light, respectively. An optical rotatory dispersion (ORD) spectrum is α plotted as a function of wavelength. Circular dichroism, on the other hand, is defined by the following equation

$$CD = \epsilon_L - \epsilon_R$$

where ϵ_L and ϵ_R are the molar absorbance coefficients for absorption of the left and right circularly polarized light, respectively.

In contrast to ORD, which is marginally useful, CD provides an important tool in empirically assessing certain elements of the secondary structure of a protein. Greenfield and Fasman (33) were the first to be reasonably successful in evaluating the secondary structure of a protein by adding, in different proportions, the CD curves for polylysine in the α helix, β conformation, and the random-coil configuration (Figure 4.7) through an iterative procedure until they matched the CD curves of unknown proteins. As a result of this study and many others over the last 15 years, it is now generally accepted that the CD spectrum of a protein can be used to predict its secondary structure. However, instead of using the different CD curves of poly(L-lysine) as standards for estimating the protein secondary structure, more accurate predictions can be obtained when the CD spectra of several proteins with known crystallographic structure are used (34).

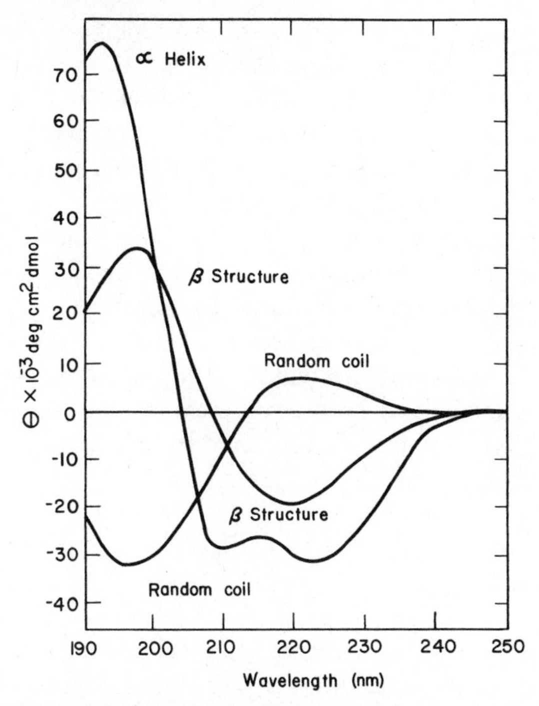

Fig. 4.7. The circular dichroism of polypeptide chains in α-helical, random-coil, and β conformations. Reprinted with permission from N. Greenfield and G. D. Fasman, *Biochemistry*, **8**, 4108–4116 (1969). Reference (33).

Analysis of the *X*-ray crystallographic structures of proteins shows that their major secondary structures can be divided into α helices, parallel and antiparallel β sheets, random regions, and β turns or bends. A β turn involves four consecutive amino acid residues where a polypeptide chain folds back upon itself to form the globular protein structure. Since different amino acids are found in these α turns, it is not surprising to find different types of turns. In fact, Chou and Fasman (35) describe 11 bend types in 29 proteins of known sequence and structure. Using these different bend types, Hennessey and Johnson (34) refined the CD method for predicting the secondary structure of a protein by including 8 types of secondary structure: helix, parallel and antiparallel β strand, types I, II, and III β turn, all remaining turns, and other structures, such as random-coil or irregular structures. After recording the CD spectra from 187 to 250 nm of 16 proteins with known *X*-ray crystallographic structures, they reconstructed the CD spectra of each in terms of the 8 types of secondary structure. The 8 types were then combined in varying proportions into 5 independent "superstructures." Hennessey and Johnson (34) emphasize, however, that the CD spectra generated from this limited set of 16 proteins will only apply to an unknown protein when its structural characteristics are well represented; proteins with an unusual complement of aromatic amino acids will not be analyzed well.

As was indicated in Chapter 1, the amino acid sequence of a protein contains the information necessary to define the secondary and the folded three-dimensional structure of a protein in solution. Consequently, considerable time and effort has been expended to develop rules and computer algorithms to predict secondary structures from amino acid sequences (36). A comparison of three different methods shows a 50 to 59% success rate in predicting the secondary structure of 62 proteins with known *X*-ray crystallographic structures (37).

FLUORESCENCE

Protein fluorescence measurements have several advantages over absorption measurements. In addition to providing a more sensitive measure of protein concentration and conformational changes, fluorescence measurements may be used to determine the fluidity of a medium and the distance between a donor and receptor chromophore (38). Both the intrinsic fluorescence of a protein molecule and the extrinsic fluorescence of an added external fluorophore can be used to study protein structure and function.

Intrinsic Fluorescence

Typical fluorescent emission spectra of the aromatic amino acids phenylalanine, tryosine, and tryptophan in aqueous solution are given in Figure 4.8. The maximum emission for phenylalanine occurs at 282 nm, tyrosine at 303 nm, and tryptophan at 348 nm. The emission spectra of phenylalanine and tyrosine are not significantly affected by changing solvent, whereas the emission maximum for tryptophan can vary from 310 to 350 nm depending on the solvent; the more nonpolar the solvent, the shorter the emission maximum. Ionized tyrosine does not fluoresce.

The emission spectra of most proteins are dominated by tryptophan fluorescence. Phenylalanine fluorescence is weak and not observed unless

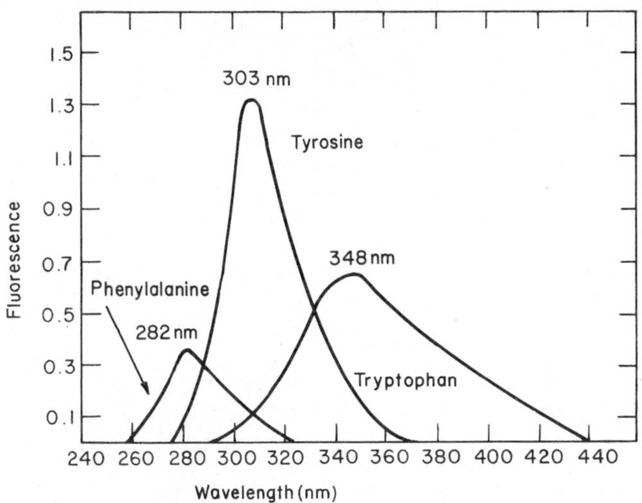

Fig. 4.8. Fluorescent spectra of the aromatic amino acids in aqueous solution at room temperature. Abscissa, wavelength (nm); ordinate, relative number of quanta. Reprinted by permission from F. W. J. Teale and G. Weber, *Biochem. J.* **65**,476–482 (1957). The Biochemical Society, London. Reference (39).

the protein contains no tryptophan and few tyrosines. Tyrosine fluorescence is usually quenched, or its absorbed energy is transferred to tryptophan and emitted as tryptophan fluorescence. The tyrosine emission of some proteins can be observed by subtracting two normalized fluorescent emission spectra of a protein, one excited at 278 nm from one excited at 295 nm; the spectra are normalized to give the same emission at 400 nm where tyrosine fluorescence is negligible. Because tryptophan fluorescence is more sensitive to the environment than tyrosine fluorescence, it provides a more sensitive measure of protein conformational changes.

Extrinsic Fluorescence

As indicated earlier, extrinsic fluorescence comes from fluorophores added to a protein. A wide variety of them are available: some are metabolites or metabolite analogs; others, known as fluorescent probes, only fluoresce when bound to proteins or other macromolecules; still other probes can be attached covalently to the protein. The structure and properties of several typical external fluorophores are given in Figure 4.9 and Table 4.2.

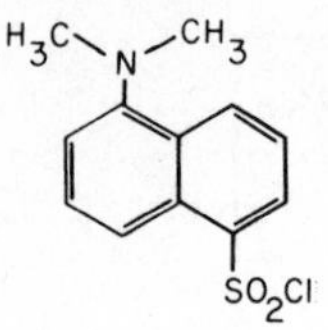

DNS-Cl

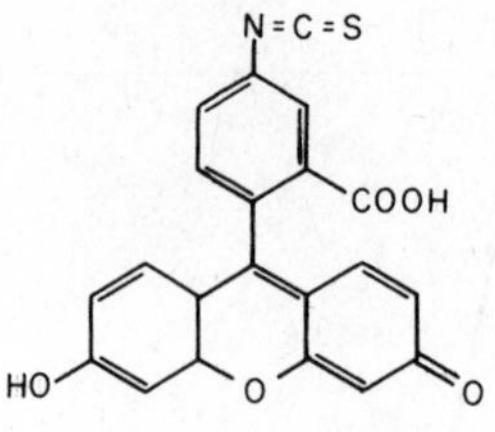

FITC

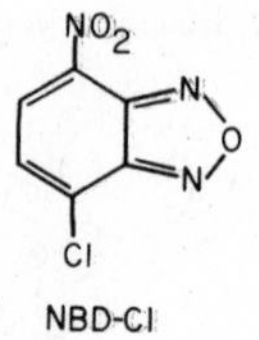

NBD-Cl

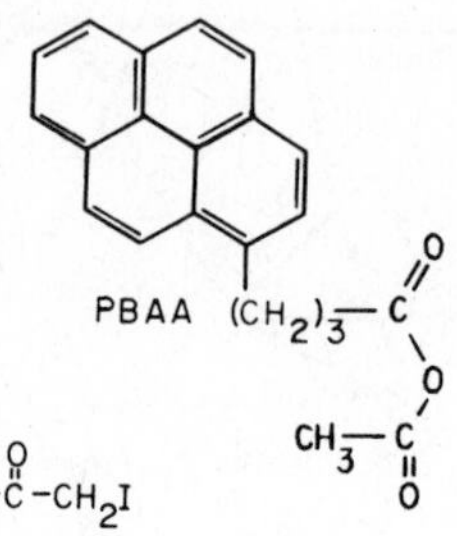

PBAA

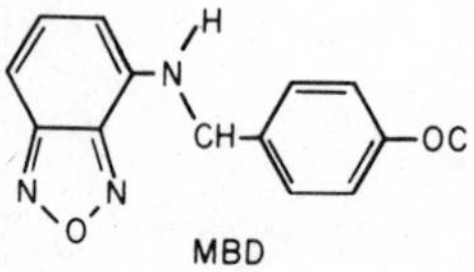

MBD

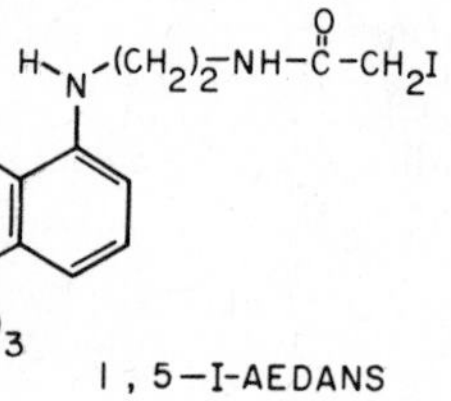

1, 5-I-AEDANS

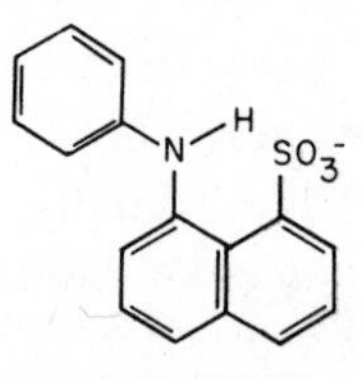

1, 8-ANS

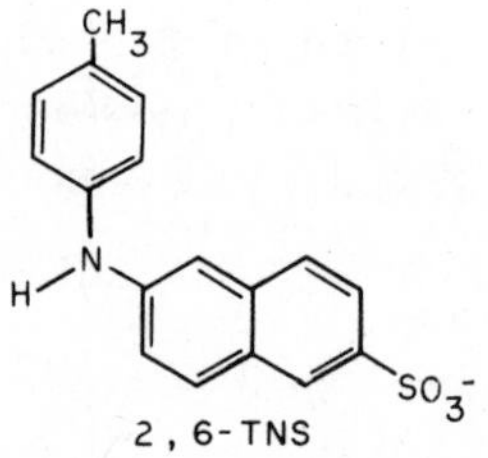

2, 6-TNS

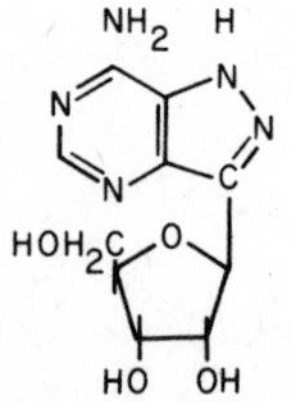

For

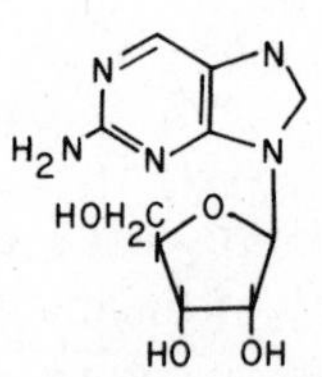

2-Amino Puo

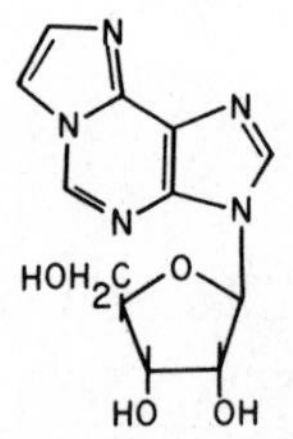

∈-Ado

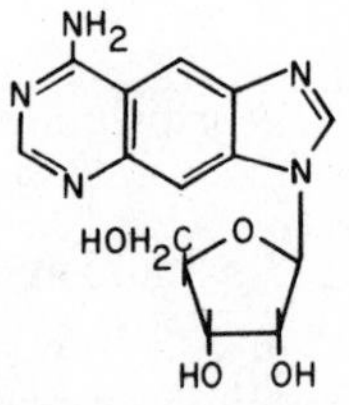

Benzo Ado

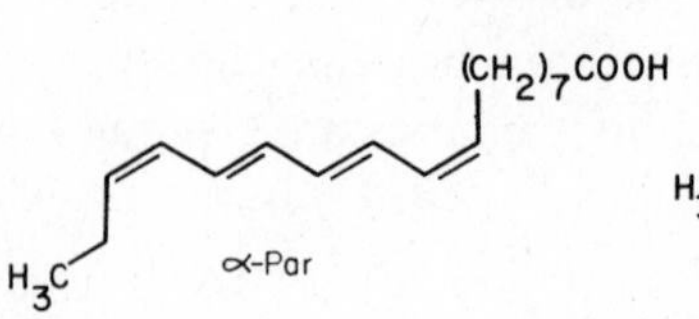

α-Par

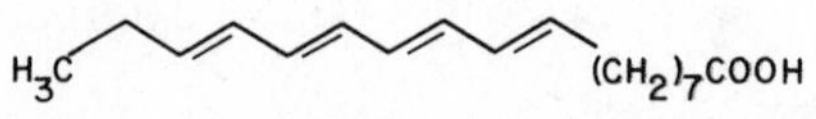

β-Par

Data from studies of protein structure and function using extrinsic fluorescence probes can be difficult to interpret. Noncovalent probes present two major difficulties. If the free (uncomplexed) probe fluoresces, its fluorescence can interfere with the fluorescence of the protein bound probe, although this problem is much less severe if high affinity probes are used. Determining the origin of the fluorescent change is the second major difficulty. For example, the change in fluorescence of a weakly bound probe may be interpreted as a change in protein conformation or it may mean that the probe has simply moved from one site to another. Fluorescent cofactors, fluorescent cofactor analogs, and covalently bound fluorescent probes are expected to have an advantage in this respect in that the fluorescent probe should remain bound to a specific region of the protein.

Fluorescent probes have many applications: they may be used in energy transfer studies to measure distances between a donor and acceptor molecule (38); to study zymogen activation; to monitor protein subunit assembly; to study the size, shape, and flexibility of macromolecules; and to detect events at an active site (51). Consult Reference (38) for a more detailed description of these methods and for further applications and references.

MOLECULAR WEIGHT

As everyone knows, the molecular weight of an enzyme is an important and useful physical parameter. It is required for calculating its turnover number, the number of ligand binding sites, its amino acid composition, and so on. Fortunately, several methods for determining molecular weight, each requiring minimal equipment and protein, are available. Equilibrium sedimentation, an exact method, and three empirical methods, sucrose density gradient sedimentation, polyacrylamide gel electro-

---

Fig. 4.9. Structures of some typical fluorescent probes for proteins. Covalent labels: DNS-Cl, Dansyl chloride; FITC, fluorescein isothiocyanate; NBD-Cl, 7-chloro-4-nitrobenzoxadiazole; PBAA, pyrenebutyrylacetic anhydride; 1,5-I-AEDANS, *N*-iodoacetyl-*N*′-(5-sulfonic-1-naphthyl)ethylenediamine. Noncovalent labels: 1,8-ANS, 1-anilinonaphthalene-8-sulfonate; 2,6-TNS; 2-toluidinonaphthalene-6-sulfonate; MBD, 7-(*p*-methoxybenzylamino)4-nitro-benz-2-oxa-1,3-diazole. Substrates (analogs of adenosine): For, Formycin; 2-AminoPuo, 2-aminopurine ribonucleoside; ϵ-Ado, ethenoadenosine; and BenzoAdo, benzoadenosine. α-Par and β-Par (α- and β-parinaric acid) are analogs of fatty acids.

Table 4.2 Spectrophotometric parameters of some useful covalent and noncovalent fluorescence probes

Fluorescence Probes[a]	Reference	Absorption Maximum in H_2O (nm)	Absorbance Coefficient ϵ $(M\ cm)^{-1} \times 10^{-3}$	Emission Maximum (nm)
DNS-Cl	40	320–340	3.4 (bound)	500–580[b]
FITC	41	495	50.3 (free)	
			42.5 (bound)	
PBAA	42	260	25	376, 395
		275	51	
		325	29	
		342	40	
1,8-ANS, Mg salt[c]	43, 44	350	4.95	436–525[b]
2,6-TNS	44, 45	223	47	413–500[b]
		263	24	
		366	4.1	
		317	18.9	
MBD	46	480	26.2	530–550[b]
ϵ-Ado	47	300 (sh)	2.6	410
		275	6.2	
		265	5.9	
		258 (sh)	4.7	
BenzoAdo	48	231	42.7	
		259	17.6	
		316	7.8	
		330	9.7	
		345	7.3	
2-AminoPuo	49	245	6.8	370
		310	7.1	
2,6-DiaminoPuo	49	255	9.2	350
		280	10	
For	49	295	9.5	340
α-Par	50	303	80	410
		318	74	
β-Par	50	299	91	

[a] See figure 4.9 for the structure of each probe.

[b] Wavelength of maximum fluorescence emission and quantum yield depends on solvent. The more polar the solvent, the longer the wavelength of maximum fluorescence and the lower the quantum yield.

phoresis, and gel permeation chromatography for determining molecular weight will be described in this section.

The majority of the molecular weight determinations reported in current literature are made using empirical methods, which, as experience has shown, give reasonable estimates of molecular weights and require less protein and less sophisticated equipment than the exact methods. On the other hand, determining the exact molecular weight by equilibrium sedimentation using the airfuge may be used more in the future; the method is simple, and many laboratories may soon have the required equipment.

In contrast to sedimentation equilibrium measurements for determining the molecular weight of a protein, which depend only on the mass of a protein, measurements made with the empirical methods noted above are affected by the shape and, in one case, by both the charge and shape of the protein. It is important, therefore, to understand the limitations of each methodology and the precautions that must be exercised, not only in the design of an experiment, but also in the interpretation of the data.

Molecular weight will be denoted by the shorthand notation M_r. It is defined as a relative molecular mass, being the ratio of the mass of a molecule to 1/12 the mass of an atom of carbon; it has no units. Some investigators prefer to use the dalton (Da) as a unit for molecular weight. The dalton is the mass of one molecule of a substance, that is, the molar mass divided by Avogadro's number (52). It is defined as 1/12 the atomic mass of carbon-12.

Sedimentation Equilibrium

The recent availability of a commercial air turbine table top centrifuge is partly responsible for the development of a convenient and exact method for determining the molecular weight of proteins (53). This method requires small amounts of protein, particularly if the protein has biological activity, or is radiolabeled. The molecular weight of ^{125}I-insulin, for instance, was determined with 87×10^{-9} mg of protein.

Protein samples in 100 μL of buffer containing bovine serum albumin are centrifuged in 175-μL untapered cellulose nitrate tubes for 24 hr; total protein concentration should be 5 mg/mL to stabilize the density gradient. After stopping the centrifuge rotor and placing the transparent tubes in a vertical position, successive 10-μL fractions are removed from the meniscus and analyzed for specific protein content by radioactivity or biological activity.

At sedimentation equilibrium

$$d \ln c/dr^2 = M_r\omega^2(1-\bar{v}\rho)/2RT$$

where c is the macromolecular concentration at the radial distince r of each successive 10-μL fraction from the center of rotation, ω the angular velocity, $\bar{v}$ the partial specific volume of the protein, ρ the solution density, and RT the gas constant and absolute temperature, respectively. The slope of a linear plot of ln c as a function of r^2 gives the molecular weight. In the molecular weight range of 6000 to 600,000, this method is reportedly accurate to ± 5%. One of the most difficult steps in this procedure is the removal of successive 10-μL aliquots from the meniscus. Pollet et al. (53) use a capillary pipette fitted to a microsyringe that has a threaded fine control screw and is stabilized and guided by a micromanipulator.

Methods for rigorously determining $\bar{v}$ of a protein are given in Reference (54). Most investigators, however, estimate $\bar{v}$ by using the following relationship

$$\bar{v} = \sum_i \bar{v}_i W_i \Big/ \sum_i W_i$$

where $\bar{v}_i$ is the specific volume of each amino acid residue and W_i the mole fraction of the molecular weight of the protein contributed by that residue. Determining $\bar{v}$ of a protein with a Kratky digital density meter (55) may become the method of choice in the future, when more laboratories obtain the instrument. Specific volumes for common amino acid residues and sugars are given in Table 4.3.

Sedimentation Velocity

Equation (4.2) expresses the sedimentation rate of a protein in solution,

$$S = M_r(1 - \bar{v}\rho)/Nf \tag{4.2}$$

where ρ is the density of the solvent, N Avogadro's number, and f the frictional coefficient of the macromolecule. The frictional coefficient f is

Table 4.3 Amino acid residue molecular weights and specific volumes[a]

Residue	Residue Molecular Wt[b]	Specific Volume of Residue (cm^3/g)
Aspartic acid	115.08	0.60
Asparagine	114.10	0.62
Threonine	101.10	0.70
Serine	87.07	0.63
Glutamic acid	129.11	0.66
Glutamine	128.13	0.67
Proline	97.11	0.76
Hydroxyproline	113.11	0.68
Glycine	57.05	0.64
Alanine	71.07	0.74
Half-cystine	102.13[c]	0.61
Valine	99.13	0.86
Methionine	131.19	0.75
Isoleucine	113.15	0.90
Leucine	113.15	0.90
Tyrosine	163.17	0.71
Phenylalanine	147.17	0.77
Lysine	128.17	0.82
Histidine	137.14	0.67
Arginine	156.19	0.70
Tryptophan	186.22	0.74
Hexoses	162.14	0.613
N-Acetylhexosamines	203.20	0.666
Fucose (6-deoxy-L-galactose)	146.14	0.678
N-Acetylneuraminic acid	291.26	0.584
N-Glycolylneuraminic acid	307.26	0.557
O-Acetyl-	59.04	0.75

[a] Taken from Reference (54) by permission.
[b] Residue molecular wt = formula molecular wt − molecular wt of H_2O.
[c] As half-cystine; the residue molecular wt of cysteine = 103.13.

dictated by the shape of the protein molecule. For a hypothetical perfect sphere,

$$f_0 = 6\pi\eta R_s$$

where η is the viscosity of the solvent and R_s is the Stokes radius of a sphere. The ratio of the experimentally measured frictional coefficient f to that calculated for an equivalent sphere, f_0, is known as the frictional ratio. For most globular proteins in dilute solution, $f/f_0 = 1.1$ to 1.3 (54).

The rate at which a protein sediments in a unit centrifugal field is called a sedimentation constant. Units for the sedimentation constant are known as Svedbergs; 1 Svedberg (S) $= 1 \times 10^{-13}$ sec. Sedimentation constants are measured in an analytical ultracentrifuge (54) or by comparison with the rate of sedimentation of one or more standard proteins (Table 4.4) in a sucrose or glycerol density gradient (72). For the latter zonal sedimentation velocity method, the protein is sedimented as a zone through a preformed density gradient, generally 5 to 20% sucrose or 9 to 36% glycerol.

Because the rate of sedimentation of an enzyme in a glycerol or sucrose gradient is relatively constant, the ratio of distances that a standard enzyme and an unknown enzyme sediment in the same gradient will be constant. Careful control of the temperature, time, centrifugation speed, and sucrose concentrations is not necessary because both the standard and the unknown are sedimented in the same gradient. For best results, the standard enzyme should have a sedimentation constant nearly equivalent to the unknown. Calculate the molecular weight of the unknown enzyme with Equation (4.3):

$$D_1/D_2 = (M_{r1}/M_{r2})^{2/3} \tag{4.3}$$

where D_1/D_2 is the ratio of the distance that the unknown enzyme sediments to the distance that the standard enzyme sediments from the meniscus.

Molecular weights obtained by zonal sedimentation velocity are in error when the partial specific volumes and the frictional ratio (shapes) of the unknown and standard enzymes differ extensively. Experience has shown that these properties of many of the soluble globular proteins do not differ sufficiently to cause a large error. It follows then that rod shaped proteins, such as myosin or tubulin, are not amenable to study by this technique unless the proper proteins are used as standards.

The zonal sedimentation velocity method for determining molecular weights is particularly useful in estimating the molecular weight of an enzyme in a crude homogenate. The positions of the unknown and standard enzymes in the gradient are located by assaying for enzyme activity. The method can be made more convenient by preparing a large number of preformed gradients and storing them frozen. The gradients are prepared as stem gradients, but each layer of the stem is frozen before the next layer is added. Gradients, which may be stored indefinitly, are thawed at 4°C at least 12 hr before use (73).

Polyacrylamide Gel Electrophoresis

Because the electrophoretic mobility of proteins in polyacrylamide gels can be determined reproducibly and accurately, only small amounts of protein (0.1–10 μg) are required, and the necessary equipment can be made in the laboratory, most investigators today use this method to determine the molecular weight of a protein. The electrophoretic mobility *m* of a protein in a polyacrylamide gel matrix is defined by Equation (4.4)

$$m = Eq/f \, \phi_r \tag{4.4}$$

where *m* is the steady state velocity of protein migration, *E* the electric field strength, *q* the charge of the protein, *f* the frictional coefficient or frictional resistance to migration of the protein, and ϕ_r, the retardation effect of the gel matrix (74). During a typical polyacrylamide gel electrophoretic experiment, the solvent medium, the gel matrix, and field strength are identical so that the mobility of a protein is proportional to q/f, the ratio of the net charge of the protein to the frictional resistance of migration. Because both the net charge and frictional resistance to migration are unknown, absolute molecular weight data cannot be obtained from a polyacrylamide gel experiment. However, as will be discussed in more detail later, the mobilities of proteins in polyacrylamide gels with differing acrylamide concentrations (pore sizes) or in gels with an increasing concentration gradient of acrylamide, can be treated so that electrophoretic data depend only on the frictional resistance of migration, that is, the size and shape of the protein; the charge of the protein is normalized. The molecular weight of the unknown can then be calculated

Table 4.4 Physical parameters of some useful protein markers

Protein	Reference	Source	Subunits		Native			
			N^a	M_r	M_r	$\overline{R}^b$ (nm)	$s_{20,w}$	Reference[c]
Trypsin inhibitor	56	Bovine pancreas	56	6,161	6,161	1.22		
Cytochrome *c*	56	Horse heart	104	11,761	11,761	1.51	1.71	57
Ribulose bisphosphate carboxylase oxygenase	58	Spinach chloroplast	116	14,346	—	—	—	
Ribonuclease A	56	Bovine pancreas	124	13,691	13,691	1.59	2.00	57
Lysozyme	59	Chicken egg white	129	14,314	14,314	1.61	1.91	57
Myoglobin	60	Horse heart	153	16,818	16,818	1.70	2.04	57
Myosin light chain A2	61	Rabbit muscle	149	15,792	—	—	—	
β-Lactoglobulin A	60	Bovine milk	162	18,363	—	—	—	
Trypsin inhibitor	62	Soybean	181	20,095	20,095	1.80	—	
Immunoglobulin G light chain	56	Human myeloma	214	23,378	—	—	—	
Trypsinogen	60	Bovine pancreas	229	23,990	23,990	1.91	2.48	57
Concanavalin A	56	Jackbean	237	25,571	102,284	3.10	—	
Chymotrypsinogen A	56	Bovine pancreas	245	25,666	25,666	1.95	2.58	57

Triose phosphate isomerase	63	Rabbit muscle	248	26,626	53,252	2.49	—	
Carbonic anhydrase B	64	Human erythrocyte	260	28,739	28,739	2.03	3.23	57
α-Tropomyosin	65	Rabbit muscle	284	32,696	—	—	—	
Lactate dehydrogenase	56	Beef heart	333	36,457	145,828	3.47	7.45	57
Aldolase	66	Rabbit muscle	361	38,994	155,976	3.55	7.35	57
Alcohol dehydrogenase	60	Horse liver	374	39,851	79,702	2.85	4.88	57
Actin	67	Rabbit muscle	374	41,719	—	—	—	
Ovalbumin	56	Chicken egg	385	42,807	42,807	2.31	3.53	68
Immunoglobulin G, heavy chain	56	Human myeloma	446	48,472	—	—	—	
Ribulose bisphosphate carboxylase/ oxygenase	69	Spinach chloroplast	461	51,154	—	—	—	
Catalase	56	Bovine liver	505	57,471	229,884	4.04	11.3	57
Albumin	56	Bovine serum	582	66,296	66,296	2.67	4.41	57
Transferrin	56	Human serum	680	76,550	76,550	2.80	5.25	70
Phosphorylase a	56	Rabbit muscle	841	97,114	388,456	4.80	13.5	71
β-Galactosidase	56	*E. coli*	1,021	116,116	464,464	5.10	15.9	57

[a] Number of amino acids in protein subunit.
[b] $\overline{R}$ is the geometrical mean radius, calculated from the molecular weight of the native enzyme assuming a partial specific volume of 0.734 cc/g.
[c] Reference for $s_{20,w}$ values.

by assuming that the standard proteins and the unknown have nearly equivalent shapes.

The molecular weight of a protein subunit can be determined by the same electrophoretic procedures as those used for native proteins, except that the protein must first be dissociated into its subunits by sodium dodecyl sulfate. Because proteins bind a nearly constant amount of this detergent per gram during this dissociation process (74), they attain a nearly constant charge-to-mass ratio and a nearly identical rod-like shape so that their relative electrophoretic mobilities are dependent only on the molecular size. Proteins with disulfide cross-links must first be reduced to allow them to assume a rod-like shape. Glycoproteins and proteins with a low pI or other proteins that do not bind the same amount of detergent per gram cannot be analyzed by this technique.

Separating proteins by electrophoresis in an increasing concentration gradient of polyacrylamide, called pore gradient gel electrophoresis, provides a better resolution of the proteins over a larger molecular weight range than electrophoresis in a uniform concentration of polyacrylamide (75). Increasing concentration gradients of acrylamide can be formed using commercial gradient makers (76) or by using a simple device described by Lorentz (77). Include 4-nitrophenol in one of the solutions of the gradient and scan the formed gels by densitometry at 405 nm to ascertain the percent acrylamide at each point in the gel (77). After electrophoresis (Figure 4.10), the log of the molecular weight of the proteins is linearly related to log percent total acrylamide at the protein position. A linear relationship exists for proteins ranging in molecular weight from 20,000 to 1×10^6 (76). The molecular weights of glycoproteins have also been successfully determined by pore gradient electrophoresis (76a).

Polyacrylamide gels of differing percentage of total acrylamide are conveniently prepared by combining different amounts of a stock solution of each component as indicated in Appendix I.

Exercise Caution. Acrylamide is a Neutrotoxin and Skin Irritant: Handle with Care; Always Use Rubber Gloves; Do Not Pipet by Mouth.

Two types of buffer systems are commonly used to electrophoretically separate proteins in polyacrylamide gels: discontinuous and continuous (homogeneous) buffer systems. The discontinuous system, originally designed by Ornstein (78) and Davis (79), is called discontinuous because different buffers are used in the electrode chambers and in the polyacrylamide gels. This system gives better resolution than the homogeneous one, which utilizes the same buffer in both electrode chambers and in the

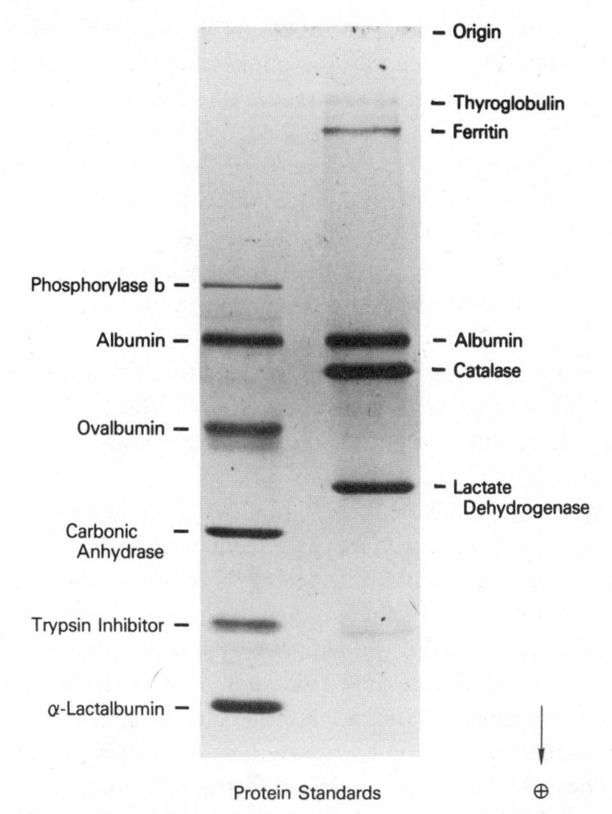

Fig. 4.10. Sodium dodecyl sulfate polyacrylamide slab gel of protein standards; linear gradient 7–25 %*T*, 1% *C*, 150 V (constant), 4 hr. Coomassie Brilliant Blue stain. Reprinted with permission from Reference (76). The photograph of the gel was graciously provided by Dr. J. F. Poduslo.

gel. Both native and sodium dodecyl sulfate dissociated proteins are separated with the discontinuous system, but sodium dodecyl sulfate dissociated proteins are often separated with the homogeneous system.

Use pure crystalline acrylamide to obtain good consistent gels with reproducible pore sizes. An aqueous solution should show a pH of 4.9–5.2. Lower pH values indicate significant contamination with acrylic acid. Degas the acrylamide solutions before adding ammonium persulfate since oxygen inhibits the initiation of polymerization by this catalyst. Do not, on the other hand, degas acrylamide solutions if they are to be polymerized photochemically with riboflavin since photopolymerization requires oxygen. After pouring the gel, overlay with water or water saturated butan-2-ol to insure formation of a flat gel enzyme solution interface. Because water saturated butan-2-ol is immiscible with water, it is easier to overlay without disturbing the surface. Note that butan-2-ol cannot be used on gels of 5% acrylamide or less because it is too dense.

For a uniform pore size from experiment to experiment, control the temperature of the exothermic polymerization reaction by conducting it in an electrophoretic chamber (with or without a water jacket) containing the electrophoresis buffer or in a cold room with good air flow. Polymerization should be complete in 10 ± 3 min. Some investigators allow the gels to stand for 24 hr before using them, but it is best not to store gels for long periods as any dehydration leads to altered pore sizes. Polyacrylamide gels prepared in a discontinuous buffer system cannot be stored in the elctrophoresis buffer; with time the ions in the buffer and the gel will diffuse into each other (80–82).

Polyacrylamide gels may be formed in glass tubes or as a slab between two glass plates. Slab gels give a more uniform electric field across several samples than tube gels. In addition, slab gels require smaller amounts of protein; the thinner the slab, the smaller the amount required. Capillary tube gels may also be used if the supply of protein is limiting (83).

When using a comparative approach such as polyacrylamide gel electrophoresis to determine an unknown molecular weight, standard molecular weight marker proteins must be used, several of which are listed in Table 4.4. These standards are stable and free of proteinases; most are available commercially. Proteins from yeast, which are often contaminated with proteinases, and multimers of bovine serum albumin, which tend to migrate faster than simple proteins (84), are not included in this listing.

Dissolve protein samples for polyacrylamide gel electrophoresis in any one of the buffer systems described in Appendix I. Each protein solution should have an ionic strength less than 0.01 *M*; dialyze to reduce it, if necessary. Glycerol is added to the sample to increase its density and viscosity so that it can be layered between the top of the gel and the electrode buffer. Use bromophenol blue as a tracking dye at alkaline pH, except in systems containing Triton X-100 and methyl green at acidic pH values (85). Bromophenol blue forms insoluble complexes with Triton.

When the molecular weight of protein subunits are determined by sodium dodecyl sulfate polyacrylamide gel electrophoresis, the protein sample conditions must be adjusted so that it binds the proper amount of detergent. The ionic strength should be low so that sodium dodecyl sulfate will form complexes with the protein rather than forming micelles. Nielson and Reynolds (74) recommend using phosphate buffer systems because the sodium dodecyl sulfate binding stoichiometry of 1.4 g/g protein was measured in this system. However, the Tris–sodium dodecyl sulfate system can be used with apparently good results (56). If the original protein sample contains a nonionic detergent, special precautions are necessary. One procedure is to add a large excess of sodium dodecyl sulfate so that it effectively competes with the nonionic detergent for the protein. If a low percentage polyacrylamide stacking gel is used above the separating gel, the mixed detergent micelles tend to concentrate at the interface between the stacking and separating gels (74). Store protein standards in small aliquots (0.1 mL) in a liguid N_2 tank or at −20°C.

Protein samples contaminated with a proteinase should be given special treatment before electrophoresis. One of several protocols may be used. One useful technique is to add the concentrated aqueous protein sample directly to a hot (boiling water bath) sodium dodecyl sulfate buffer containing 2-mercaptoethanol and allow it to remain for 2–3 min. Extensive boiling (20 min) will cleave peptide bonds (86). If this treatment does not prevent proteolytic degradation, preincubate the protein with 1-m*M* phenylmethylsulfonyl fluoride or another proteinase inhibitor (see Table 3.1) and then disperse rapidly in a strong denaturant such as a solution containing guanidine thiocyanate, 2-mercaptoethanol, and ethylenediaminetetraacetic acid (EDTA) (see Chapter 1). The denaturant can then be removed by dialysis, first against urea and finally against sodium dodecyl sulfate. Removing guanidine hydrochloride prevents formation of the insoluble guanidine dodecyl sulfate.

Detecting Proteins after Electrophoresis

A variety of specific and nonspecific methods of differing sensitivities are available for detecting proteins in polyacrylamide gels or other electrophoretic support media. Specific methods are based on ligand binding reactions, antigen–antibody reactions, and enzyme activities. Proteins with cofactors or posttranslational modifications may also be detected specifically. Most investigators stain proteins nonspecifically with dyes or with the more sensitive silver stain. Other more sensitive detection methods use proteins labeled before electrophoresis with radioactive, fluorescent, or chromophoric compounds. Unstained polyacrylamide gels can be scanned at 280 nm or at other appropriate wavelengths; gels prepared in quartz tubes can be scanned directly. Alternatively, proteins in unstained gels can be visualized directly by UV induced fluorescence (87), or proteins in urea gels can be visualized by placing them in distilled water after electrophoresis (88). Presumably denatured protein precipitates in the gel as urea diffuses out; maximum contrast occurs in 90 min. Immunochemical methods can also be used to detect proteins nonspecifically. For example, proteins from electrophoretograms can be electroeluted onto an immobilizing matrix such as nitrocellulose or Zeta-bind paper and derivatized with 2,4-dinitrofluorobenzene (89) or pyridoxal phosphate and sodium borohydride (90), then incubated with antibodies specific either for the dinitrophenyl or the phosphopyridoxyl group, and subsequently visualized in a variety of ways.

Two Coomassie Brilliant Blue dyes are commonly used in enzymological laboratories for staining polyacrylamide gels: Coomassie Brilliant Blue G-250 (CI 42,655) and Coomassie Brilliant Blue R-250 (CI 42,660); the suffixes G and R signify greenish-blue and reddish-blue hue, respectively, while the suffix 250 means that the sample contains 2.5 times as much dye as the standard supply (91). Coomassie Brilliant Blue dyes are more sensitive than other dyes and, therefore, are the stain of choice for most applications, although some investigators prefer to use amido black (92) or fast green FCF (93). Since most commercial samples of these dyes contain neutral salts and intermediates or by products of the synthetic process, special efforts should be made to standardize the concentration of dye in each staining solution (91), particularly when quantitating proteins in the gel.

Silver is 100 times more sensitive than Coomassie Brilliant Blue in staining proteins in polyacrylamide gels (94–97). Furthermore, under certain prescribed conditions, silver staining produces colored spots, which can be used to identify specific proteins (98). Gels first stained with Coomassie Brilliant Blue can also be stained with silver or vice versa (95). However, as emphasized during a workshop on silver staining (99), there are problems associated with this method, including protein fixation, loss of sensitivity, background staining, and destaining.

Before staining proteins in gels with silver, many investigators fix them in the gels with 50% methanol plus 10% acetic acid; for ultrathin gels (0.5 mm), 20% trichloroacetic acid is recommended because the methanol–acetic acid solution washes protein out of these gels. Aldehydes such as glutaraldehyde or formaldehyde are often used in a second fixation step. To obtain a clear background, it is important to wash the gel extensively after the aldehyde fixation to remove glycine, aldehydes, ampholytes (if present), and sodium dodecyl sulfate. For 1.5-mm gels, 18–24-hr wash times at 65°C are required. Placing the gel in 70% (w/v) poly(ethylene glycol) (PEG)-2000 enhances the rate of removal of sodium dodecyl sulfate (100). The concentrated PEG shrinks the gel, creating an efflux of buffer and sodium dodecyl sulfate from it. Because silver staining is very sensitive to temperature fluctuations, staining at 50°C is recommended where a 3 to 5°C temperature fluctuation has little significance.

Staining and destaining procedures for both Coomassie Brilliant Blue dyes, amido black, and fast green FCF are given in Inset 4.1. References (80–82, 91, 92, and 101) provide details for staining proteins with other dyes. Because silver staining methods are still in their development phase, additional details for their use are not provided in Inset 4.1.

Labeling proteins with radioactive (107–110), fluorescent (111), or chromophoric (112) compounds prior to electrophoresis should not influence their electrophoretic mobility. Labels which do not are iodine-125 (107), tritium (108, 109), and ^{14}C sucrose (110). Proteins labeled with fluorescamine (111) or 4-dimethylaminoazobenzene-4′-sulfonyl chloride (dabsyl chloride) (112) have an advantage in that the electrophoetic migration of the labeled proteins can be monitored continuously. However, fluorescamine introduces a net negative charge and, therefore, should only be used in sodium dodecyl sulfate polyacrylamide gels. Because the dabsylated proteins are more hydrophobic, they must be used with solvent systems in which they are soluble. This latter derivative can also be eluted from the gels, hydrolyzed, and analyzed for their N-terminal amino acids (112).

Most methods for detecting specific enzymes or proteins on electropherograms, even electropherograms of crude extracts, utilize the catalytic activity of the enzyme. Both native and sodium dodecyl sulfate dissociated proteins can be detected in situ, or the gels can be sliced into thin sections from which the enzyme can be leached out with buffer, renatured if necessary, and assayed for activity (1, 113) (see Chapter 1 for details on renaturation in situ). Nicotinamide adenine dinucleotide or $NADP^+$ dependent dehydrogenases are detected by noting the appearance of the fluorescent reduced nucleotide product directly (114), or by using phenazine methosulfate to facilitate the reduction of nitroblue tetrazolium by the reduced nucleotides to form the insoluble formazan (115). Be careful though, as both of the latter procedures are plagued by artifacts (115). One of the artifacts, often called nothing dehydrogenase, is probably

INSET 4.1.

Procedures for Staining Proteins after Electrophoresis in Polyacrylamide Gels

---

Coomassie Brilliant Blue G-250

To stain native proteins in polyacrylamide gels, place gels into 0.04% (w/v) solution of Coomassie Brilliant Blue G-250 in 3.5% (w/v) perchloric acid for 90 min. Then transfer the gels to 5% (w/v) acetic acid for destaining and storage (91). Prepare the dye solution, which is stable indefinitely at room temperature, by stirring the dye and perchloric acid for 1 hr at room temperature; then filter through a Whatman No. 1 paper, followed by a 0.45-μm membrane filter.

Coomassie Brilliant Blue R-250

Coomassie Brilliant Blue R-250 is normally used to stain sodium dodecyl sulfate polyacrylamide gels. Place gels in a staining solution containing 1.25 g Coomassie Brilliant Blue R-250, 227 mL methanol, 46 mL glacial acetic acid, and 500 mL H_2O. Dissolve the dye in methanol before adding acetic acid and water; then filter through a Whatman No. 1 filter paper. Allow gel to remain in staining solution for 2–12 hr at room temperature. Destain gels in a solution containing 50-mL methanol, 75 mL acetic acid, and 1 L water. Adding a mixed bed ion exchange resin or charcoal in a suitable device enhances the rate of destaining (102, 103). Store in the dark in 7% acetic acid.

For rapid fixation and staining of native proteins, Maurer and Allen (104) recommend immersing gels in 12.5% trichloroacetic acid at 65°C for 30 min to fix proteins. The gels should then be rinsed with tap water and stained at 65°C for 30 min in 0.2% Coomassie Brilliant Blue R-250 in a 45:45:10 by volume mixture of water, absolute ethanol, and acetic acid. As noted above, destain gels at room temperature. Store Coomassie Brilliant Blue R-250 stained gels in the dark in 7% acetic acid.

INSET 4.1. (*Continued*)

Amido Black

Native proteins in polyacrylamide gels may be stained overnight in 0.1% amido black in 7% acetic acid. Staining will be more rapid if the mixture is heated at 95°C for 10 min. Destain gels by washing repeatedly in 7% acetic acid or in water:methanol:acetic acid (5:5:1). Store gels in 7% acetic acid. Adding 1 mg/L of amido black to the 7% acetic acid storage solution prevents the slow loss of dye from the stained protein. For long term storage, keep in the dark in closed vessels (92).

Fast Green FCF

Stain polyacrylamide gels overnight in 1% dye in 7% acetic acid followed by destaining in 7% acetic acid. For long term storage, place in the dark in 7% acetic acid containing 1 mg/L dye. Fast green FCF does not bind ampholytes so it is useful for staining isoelectric focusing gels (93).

The sensitivity of the methods for detecting proteins in polyacrylamide gels can be enhanced four- to six-fold simply by reducing the size of the stained gel by placing it in concentrated PEG-6000 for varying times. This procedure may eliminate the need to concentrate a protein sample before it is subjected to electrophoresis (105).

Peptides with molecular weight less than 5000 often diffuse out of the gels during the staining and destaining procedures. Incubating the gels after electrophoresis with 5% formaldehyde in water for 1 hr before initiating the staining process prevents this loss (106).

due to alcohol dehydrogenase and alcohol contamination in the reagents; lyophilizing all reagents prior to staining should eliminate it (115). Many other specific enzyme reactions may be detected using procedures described in References (116) and (117); enzymes with bound cofactors that absorb light at unique wavelengths can be detected by scanning the gel in a suitable device with a spectrophotometer (118).

Other methods for detecting specific proteins in electropherograms take advantage of the specificities of antibodies and other affinity probes, such as lectins. To detect an antigen, overlay the gel with a cellulose acetate membrane soaked in the appropriate antiserum (119). If cellulose acetate strips are used as the electrophoretic support medium, soak strips with 0.5 mL antiserum previously diluted 1:5 in phosphate buffered saline containing 40 g/L PEG-6000 for 5 min. Alternatively, the affinity probe (antigen antibody or lectin) can be covalently coupled to diazonium groups on an activated cellulose filter and then overlaid on the gel containing the separated proteins (120). Proteins that interact with the affinity probe then diffuse from the gel to the filter. Instead of transferring the protein from the polyacrylamide gel to the activated cellulose by simple passive diffusion, the proteins can be transferred electrophoretically to nitrocellulose (121, 122), an electrophoretic method of transfer known as "Western blotting." The immunofixed proteins may be stained by Coomassie Brilliant Blue or amido black. Radioautography, after adding radiolabeled Staphylococcus protein A (121), lectins (123), or nucleic acids (124) detects specific binding proteins; lectins or antibodies may also be labeled with enzymes (125, 126) prior to use.

As staining technology becomes more sophisticated and methods for detecting proteins more sensitive, artifacts, probably due to contaminating proteins, become more common. Several investigators have reported artifacts at two spots equivalent to M_r = 54,000 and 68,000 (99, 127, 128). These artifactual proteins were often attributed to the use of 2-mercaptoethanol (99, 127), but a recent report indicates that the contaminating proteins are skin proteins, especially keratins, with M_r = 54,000 to 57,000 and 65,000 to 68,000. Because these proteins appear to contaminate solutions, reagents, and equipment, special precautions are necessary when sensitive staining methods are employed.

Measuring Mobility

When measuring the migration distance of both the electrophoretic front and the stained protein band, remember that polyacrylamide gels both shrink and swell during staining and destaining. For careful work, measure the length of the gel and the migration distance of the electrophoretic front before removing the gel from its mold. Some investigators (129, 130) photograph the gel after completing the electrophoresis and then calculate the relative mobility with Equation (4.5):

$$m_r = (x_2/x_1) \times (l_1/l_2) \tag{4.5}$$

where x_1 is the position of tracking dye in unstained gel; x_2 is the position of protein in stained gel; l_1 is the length of unstained gel; and l_2 is the length of stained gel. For best results, measure the migration distances and the gel length to 0.1-mm accuracy by scanning the stained gel with a densitometer.

After removing the gel from its mold, mark the electrophoretic front by inserting a small wire, or pierce the gel with a small wire previously dipped in India ink. Make sure to note whether or not the stacking gel becomes unstacked, that is, does it migrate with the true electrophoretic front identified as a small band with a different refractive index (Schlieren line) in the vicinity of the dye front (56). Locating the true electrophoretic front is critical because a small error in identifying the position of the front will produce a large error, particularly in the relative mobilities of large molecules, resulting in curved Ferguson's plots.

Analyzing Electrophoretic Data

As noted earlier [Equation (4.4)], the electrophoretic mobility of a protein in a gel is a function of the electric field strength, the charge of the protein, the frictional coefficient or frictional resistance to migration, and the retardation effect of the gel matrix. Like the sedimentation constant, the electrophoretic mobility is affected by both the size and the shape of the protein. However, the frictional resistance to migration in polyacrylamide gels is more complex than the frictional coefficient of a protein in free solution since the pore sizes determined by the concentration of the gel affect the electrophoretic migration.

There is considerable evidence (131) to show that the log of the relative electrophoretic mobility of a protein m_r is linearly related to the total gel concentration T through Equation (4.6):

$$\log m_r = \log m_{r0} - K_R T \tag{4.6}$$

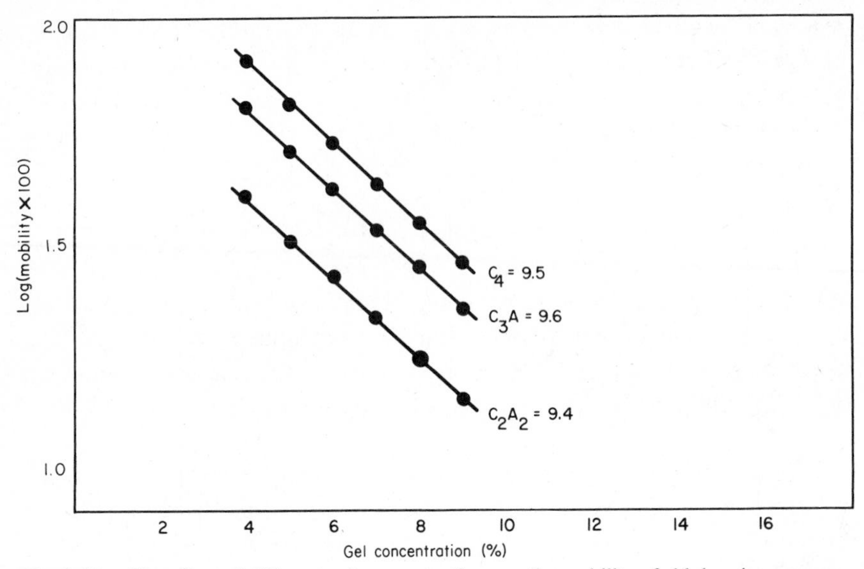

Fig. 4.11. The effect of different gel concentrations on the mobility of aldolase isoenzymes. The negative slopes of the lines are noted on the figure. Reprinted with permission from Reference (133).

where m_{r0} is the relative electrophoretic mobility at 0% gel and K_R, the retardation coefficient, is the slope of a plot of log m_r versus $\%T$. Called the Ferguson plot (132) it was first applied to the determination of the molecular weight of native proteins by Hedrick and Smith (133) using polyacrylamide gel electrophoresis. The retardation coefficient K_R is affected by both the Stoke's radius of the protein and the percent of N,N'-methylenebisacrylamide in the gel; if the ratio of bisacrylamide to acrylamide is kept constant, a plot of $\sqrt{K_R}$ versus $\bar{R}$ or log K_R versus log M_r will give a linear standard curve (131, 133). Keep in mind though, for these functions to be linear, all standard proteins and unknowns must have similar shapes. This problem will be discussed in more detail later.

Figure 4.11 is a plot of Equation (4.6) showing the log of the mobilities of three isozymes of aldolase as a function of the percent total gel concentration $\%T$.

$$T = \frac{(\text{acrylamide} + N,N'\text{-methylenebisacrylamide}) \text{ in g} \times 100}{\text{volume in mL}}$$

The projected intercepts of the three lines in Figure 4.11 to $T = 0$ are

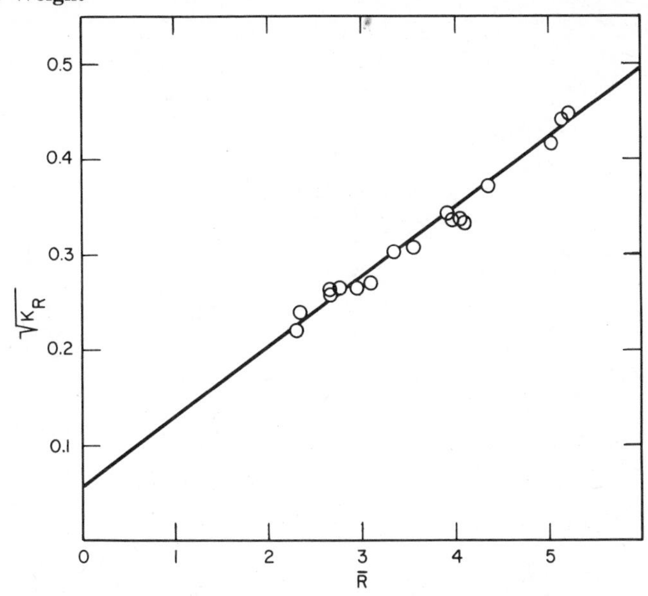

Fig. 4.12. Relationship between retardation coefficient (K_R) and geometrical mean radius $\overline{R}$. Data are from Table 2 of Reference (133).

different, but their K_R values are identical. This behavior indicates that the size and shape of these three isozymes are the same within experimental error, but their charge per unit mass is different. The K_R values, therefore, are independent of the charge of the protein, and a plot of $\sqrt{K_R}$ for several standard proteins as a function of their geometrical mean radii, $\overline{R}$ (Figure 4.12), gives a standard curve for determining the molecular weight of an unknown. $\overline{R}$ is calculated assuming a partial specific volume of 0.734 cc/g and no hydration. The value for the partial specific volume is obtained from the specific volume of each amino acid (Table 4.3) and the average amino acid composition of 184 sequenced proteins (134).

To determine the molecular weight of protein subunits by polyacrylamide gel electrophoresis, first dissociate proteins with sodium dodecyl sulfate. As indicated earlier, under the proper conditions, proteins bind a stoichiometric amount of the negatively charged detergent to form rigid rod shaped complexes with a major axis length of 0.074 nm/amino acid (135). The excess negative charge of the bound detergent swamps out the intrinsic charge of the protein subunit so that all polypeptides have essentially the same charge-to-mass ratio and electrophoretic mobilities are proportional to their molecular weight or to the number N of amino acids

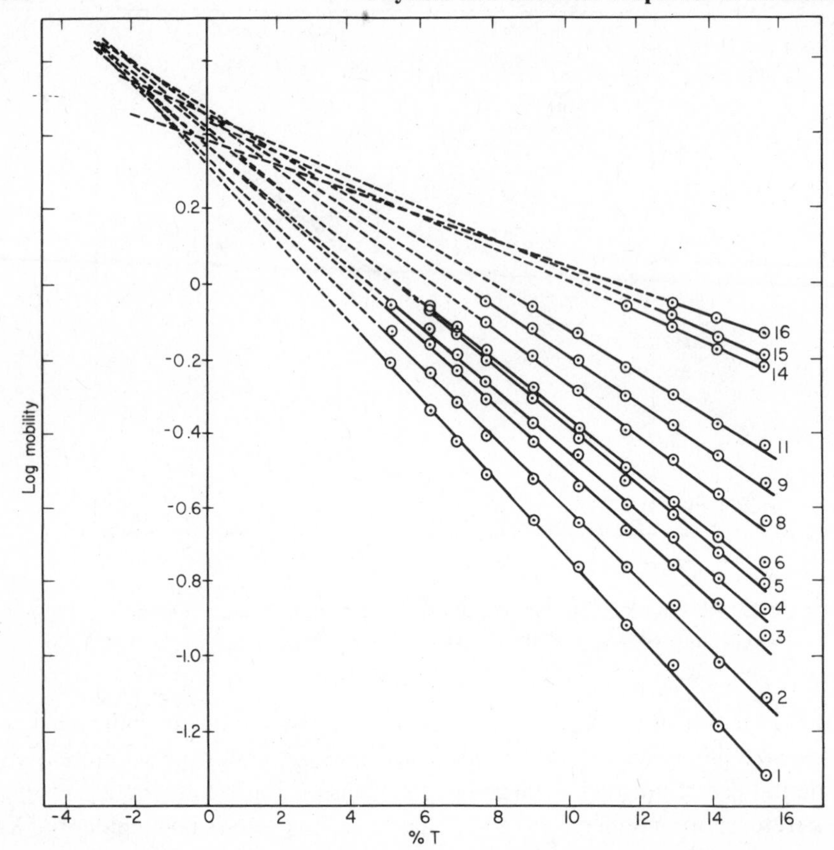

Fig. 4.13. Ferguson plot of the log mobility of several molecular weight markers as a function of $\%T$. The numbers correspond to the proteins listed in Table 1 of Reference (56). Because of crowding, data for proteins numbered 7, 10, 12, and 13 are not included. The relative electrophoretic mobilities of each protein at each $\%T$ were provided by Dr. Gary L. Peterson, Department of Biochemistry and Biophysics, Corvallis, Oregon 97331.

in the polypeptide chain. Peterson and Hokin (56) find N to be a more precise calibration parameter. (Table 4.4 gives N for several protein molecular weight markers.)

Figure 4.13 is a Ferguson plot of the log mobility versus $\%T$ for several standard marker proteins in sodium dodecyl sulfate polyacrylamide gels. Note that all lines, except those for cytochrome *c* and bovine trypsin inhibitor, tend to extrapolate back to a common point as should be the case, if all protein subunits have the same charge-to-mass ratio. As Sandermann and Strominger show [see Figure 2 of Reference (136)], com-

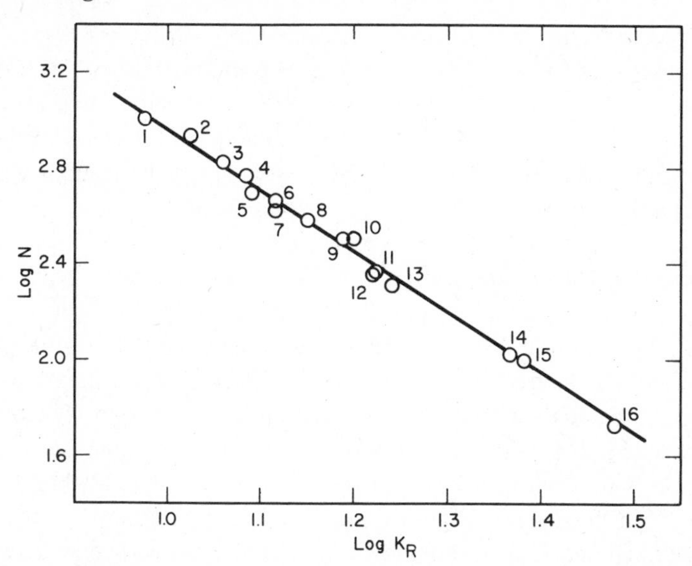

Fig. 4.14. A presentation of log N as a function of Log K_R. The values for K_R, the retardation coefficients, are obtained from the slopes of the lines in Figure 4.13. The numbers on the graph correspond to the marker proteins 1 through 16 given in Table 1 of Reference (56).

plexes of sodium dodecyl sulfate with polypeptides of M_r less than 13,000 no longer behave as the larger molecular weight protein markers. For these short polypeptides, the length of the minor axis begins to approximate the length of the major axis, and therefore, they migrate as spheres instead of as rod shaped particles (135).

The molecular weight of an unknown protein is obtained from a standard curve prepared by plotting log K_R, that is, the log of the slopes of lines in figure 4.13, versus log N, as shown in Figure 4.14. The number of amino acids in the unknown is obtained from the standard curve. To convert N to molecular weight, multiply by 112, the weight average molecular weight of amino acids in 184 sequenced proteins (134).

As a shortcut to the above procedure, many investigators plot log M_r versus relative mobility at one fixed gel concentration and fit the data to a straight line. However, as noted in Figure 1 of Reference (56), such a plot is only linear over a limited molecular weight range. For a more extensive molecular weight range, the plot gives a sigmoid curve, which can be fitted by an iterative least squares method to Equation (4.7)

$$\log m_r = \log a - b \log(1 + M_r/c) \tag{4.7}$$

where a, b, and c are regression constants (56). On the other hand, linear calibration curves of log M_r versus relative mobility for proteins with M_r ranging from 2500 to 90,000 are obtained with a discontinous sodium dodecyl sulfate polyacrylamide gel system using special buffers containing 8 M urea in the separating gel (137). The system may be particularly useful for determining the molecular weight of small polypeptides.

Gel Permeation Chromatography

Because proteins with different molecular weights require different volumes to elute them from gel permeation columns, gel permeation chromatography can be used to determine their molecular weight. The method can be applied to a wide range of different sized proteins because gel permeation media with different porosities are available (Table 3.7). In addition, high performance gel permeation chromatography makes this technique particularly attractive (138, 139). The method is easy to apply, small amounts of protein are required, and impure preparations or crude extracts can be used.

Separating different size molecules by gel permeation chromatography is based on the partitioning of a solute molecule by diffusion between the solvent in the interior of the gel matrix and that in the exterior. Small molecules that freely diffuse between these spaces are eluted in the total bed volume V_t; large molecules sterically prohibited from entering the interior of the gel matrix are eluted in the void volume V_0, as diagrammed schematically in Figure 4.15. This figure also shows the elution profile of an intermediate sized molecule. The volume required to elute molecules in this the sizing range is called the elution volume V_e. The internal volume of the matrix V_i, is defined as

$$V_i = V_t - V_0 \tag{4.8}$$

The fraction of the internal volume accessible to a macromolecule is called a distribution coefficient or partition coefficient K_p and is defined as

$$K_p = (V_e - V_0)/(V_t - V_0) \tag{4.9}$$

Partition coefficients express the fundamental sizing property of a gel permeation matrix. In contrast to the measured elution volume of a com-

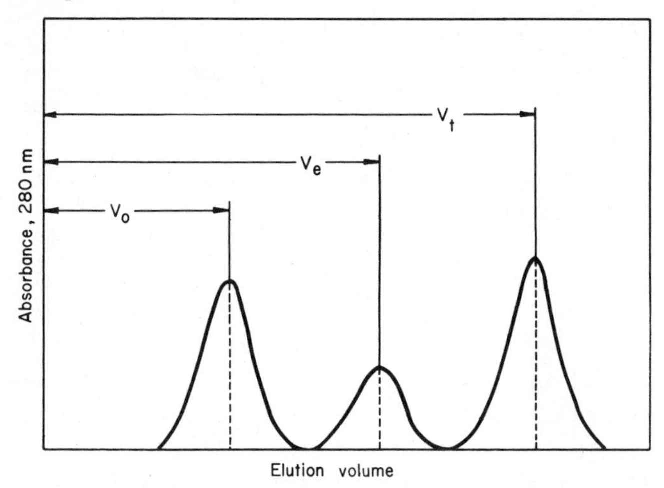

Fig. 4.15. Hypothetical elution profile of a gel permeation column.

pound from a particular column, which cannot be compared directly to any other quantity except another elution volume of the same column, partition coefficients are the same from one column to another of a given gel permeation matrix. A close examination of Equation (4.9) shows that molecules totally excluded from the gel matrix have a $K_p = 0$; molecules that freely diffuse between the inner and outer spaces of the gel matrix have a $K_p = 1$. If molecules have a K_p greater than one, they are retarded by the column because of some interaction with the matrix.

To determine K_p, one needs to know the void volume V_0, the internal volume V_i, and finally the volume required to elute standard molecular weight markers and the unknown from the column. Experimentally V_0 is the volume required to elute a molecule that is totally excluded from the matrix, such as dextran blue or DNA. Large proteins, like porcine thyroglobulin (M_r = 670,000) or bovine glutamate dehydrogenase (M_r = 998,000), have also been used (139). Blue dextran is not suitable for defining V_0 in high pressure gel permeation columns as it adsorbs to the matrix (139). The total volume V_t is that required to elute molecules that freely diffuse through the matrix, such as dithiothreitol (140), glycyltyrosine (141), dinitrophenylalanine (142), and tritiated water (143).

The elution parameters V_0, V_e, and V_i should be determined as precisely as possible. Collect the eluent after carefully layering the sample under the solvent onto the top of the gel permeation media. Sucrose (20%) or glycerol (25%) increases both the density and the viscosity of the sample and stabilizes the solution while it is layered onto the column bed. Because of the design of high pressure chromatographic systems, applying a sample and measuring its elution parameters is more precisely controlled.

Most elutions are run at constant flow rate so that the elution volumes can be obtained from the volume in each fraction multiplied by the fraction number containing the maximum concentration of the solute. For maximum precision, Fish et al. (142) find it desirable to use elution weight instead of elution volume; weigh test tubes prior to and after collection of eluted solvent to determine elution weight.

A number of methods are available for analyzing gel permeation data to obtain the molecular weight of an unknown protein (142, 144–147). The most popular method uses a plot of the log M_r versus K_p, the partition coefficient (147). I prefer the method proposed by Ackers (146), who argues that K_p—the fraction of the internal volume that is penetrable by a solute molecule of radius R—can be represented by a Gaussian probability curve. The partition coefficient K_p varies from 0 to 1 and is described by the error function complement (erf^{-1}) of the Gaussian distribution. Equation (4.10) describes the relationship between K_p and the molecular radius of the protein molecule R.

$$R = a + b\ \underline{erf}^{-1}(1-K_p) \tag{4.10}$$

A plot of the geometric mean radius, $\overline{R}$, against $\mathrm{erf}^{-1}(1-K_p)$ gives a straight line with slope a and intercept b (Figure 4.16); a and b are calibration constants for a given gel column. Appendix II provides $\mathrm{erf}^{-1}(1-K_p)$ values.

The reliability of a molecular weight determined by gel permeation chromatography depends on the accuracy with which the partition coefficients have been determined and their interpretation. In general, K_p values can be determined reproducibly and accurately; the larger the K_p values, the greater the accuracy. The interpretation of K_p, on the other hand, is more troublesome. Since K_p values are governed by steric constraints that exclude the solute molecules from the internal space, they are—like the retardation coefficients obtained by polyacrylamide gel electrophoresis—profoundly affected by the shape of the molecules. Perfect spheres will have different K_p values than rod shaped particles of the same molecular weight. Concentration dependent protein association–dissociation reactions will also affect the K_p values.

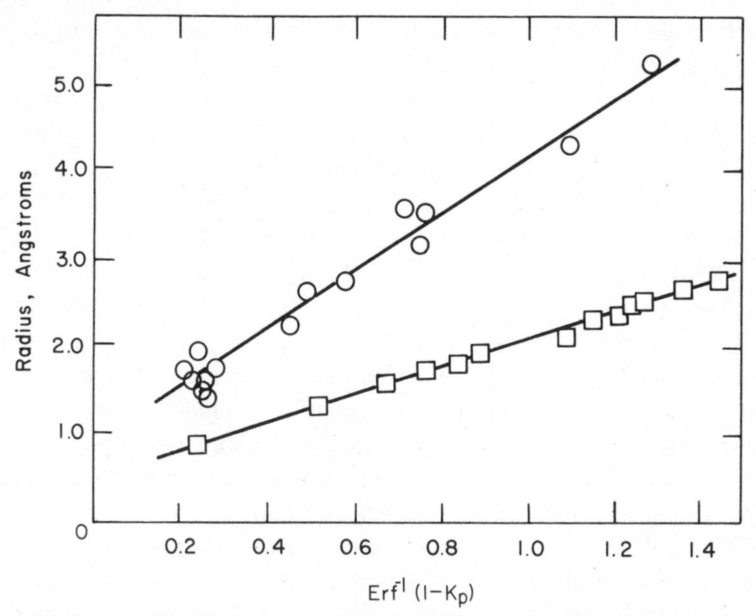

Fig. 4.16. Presentation of the geometric mean radius as a function of the error function complement of K_p for several standard proteins. The data for the top line (○) from Reference (141) are for native proteins eluted from a gel permeation column using high pressure liquid chromatographic techniques. Data for the lower line (□) from Reference (142) are for proteins eluted from 6% agarose in 6 *M* guanidine·HCl, 0.1 *M* 2-mercaptoethanol. Appendix II gives the error function complement applied to gel permeation chromatographic data.

SUBUNIT COMPOSITION

Number of Subunits

Many proteins, particularly those subject to metabolic regulation, function as multimers of more than one polypeptide. To determine the number of polypeptides in a multisubunit protein divide the overall molecular weight by the molecular weight of the subunit (148). As long as each polypeptide has the same molecular weight and the number of polypeptides does not exceed four, this approach is reasonably accurate. However, if the protein has more than four subunits, the precise number may be difficult to determine because the molecular weight of the protein and its subunits cannot be determined with sufficient accuracy. For example, a protein with M_r = 160,000, if composed of four subunits, would have a subunit M_r = 40,000; for five subunits, M_r = 32,000; and for six subunits, M_r = 26,670. Thus an error of ±10% in determining molecular weights makes it increasingly difficult to determine the number of subunits as their numbers increase.

The number of subunits in a protein may be confirmed by: (a) partially cross-linking the protein and separating the resulting cross-linked peptides by sodium dodecyl sulfate polyacrylamide gel electrophoresis and (b) electrophoretically resolving the isozymes of a protein created by the dissociation and reassociation of a 1:1 mixture of a native protein and one in which its subunits have been chemically modified. A partially cross-linked multisubunit protein will contain some protein molecules with no cross-links, some with one cross-link, some with two cross-links, some with three, and so forth until all subunits are cross-linked together. When such a cross-linked tetrameric protein is subjected to sodium dodecyl sulfate polyacrylamide gel electrophoresis, the number of bands on the gel will equal the number of subunits, if one assumes that the polypeptide with three and four cross-links migrates identically.

Alternatively, a protein solution can first be divided in two parts. One of the two parts is chemically modified to change its charge characteristics [e.g. by maleylation (149)]; the two parts are then recombined, dissociated into subunits by guanidine·HCl, reassociated by reducing the concentration of the denaturant, and resolved by polyacrylamide gel electrophoresis. The number of bands (artificial isozymes) appearing on the gel will be the number of subunits plus one.

Another technique for estimating the number of subunits is to determine the number of ligand binding or catalytic sites. The number of sites is often equivalent to the number of subunits (149), but because some monomeric enzymes contain more than one catalytic site (150), this method cannot be used as a reliable confirmation of the number of subunits.

Are the Subunits Identical?

Several methods may be used to determine whether the subunits in a multisubunit protein are identical or not. First, isozymes indicate the existence of nonidentical subunits; for example, lactate dehydrogenase composed of H and M polypeptide chains reassociates after dissociation in guanidine·HCl into five isozymes: H_4, H_3M, H_2M_2, HM_3, and M_4. However, it is important to confirm these results with other methods because some proteins with different subunits do not reassociate to form isozymes. Tetrameric hemoglobin composed of A and B chains, for example, reassociates after dissociation in guanidine·HCl to form the tetramer A_2B_2. Separating the subunits using chromatography (151–153) or isoelectric focusing (154) under denaturing conditions and determining their amino acid composition confirms their nonidentity. As will be discussed in more detail later, peptide maps and C- and N-terminal amino acids, although not as convincing as peptide maps, also provide evidence for determining whether the subunits of a protein are identical.

AMINO ACID COMPOSITION AND SEQUENCE

The amino acid composition of a protein indicates its complexity, describes some of its properties, and provides data for calculating its partial specific volume, (see Table 4.3), and exact molecular weight. The sequence of amino acids and the cross-links between different amino acids in a protein define its primary structure and, for a number of reasons, is useful to know. As already discussed in this chapter, a sequence can be used to predict the secondary structure of a protein. Identifying a portion of an amino acid sequence, that is conserved during evolution, suggests an important role in the function of the protein. The sequences also provide clues as to how the protein evolved (155). Finally, an amino acid sequence, or portions of a sequence, can be used to prepare complementary DNA (cDNA) probes for cloning genes to be used for the biosynthesis of the protein. Some of the basic essentials for determining the composition and sequence of a protein is described in this section: Most of it is abstracted from References (156)–(158).

Amount of Protein Required

The amount of protein required to determine the composition and/or sequence of a peptide depends on the sensitivity of the available instrumentation and methodology. When special precautions are used in preparing buffers, particularly to remove ammonia from them, 1–40 nmol of protein is sufficient to obtain amino acid composition data (159). By carrying synthetic amino acid mixtures through the hydrolysis procedure to correct for contamination and losses, it is possible to determine composition data with 1 to 10 pmol of protein (160, 161). The amount of protein required for determining its composition, therefore, is not limited by available methodology but rather by the contaminating amino acids in the buffers, the derivatizing reagents, and the acids used for hydrolysis (162). Buffers are often contaminated with ammonia and other compounds, presumably amines, which give rise to artifactual peaks. Commercial hydrochloric acid contains amino acids. Glassware must be clean. Gloves should also be worn because fingerprints are a source of many contaminants.

Factors that influence the amount of protein required for sequencing are (a) the size of the protein, (b) the distribution of amino acids in the sequence, (c) unusual amino acids in the protein, and (d) the purity of the reagents. The longer the polypeptide chain, the more material required; longer polypeptide chains require more processing before the sequencing can be started. Each processing step, particularly those required to separate and purify peptides, results in losses of peptides. Additional protein is required if the distribution of susceptible amino acids in the sequence, where cleavage occurs, results in large peptides, requiring, in turn, additional peptides before sequencing.

As Walsh et al. (158) points out, there appears to be an inverse correlation between the availability of a protein or peptide and the interest in its amino acid sequence, provoking leading investigators to develop more sensitive methodologies. One such methodology, the spinning cup sequenator, is now used routinely at the 10–50-nmol level (163, 164). A modified sequenator employing gas phase reagents instead of liquid phase reagents in the Edman degradation is used at the 10–100-nmol level (165, 166); it can be used with as little as 5–50 pmol.

Purity of Protein

A protein should be 95% pure on a weight basis before its composition or structure can be determined; no single polypeptide should be present in an amount exceeding 3%. Also, before beginning an analysis, remove coenzymes and cofactors (159). Nonidentical subunits in a multimeric protein are first separated and purified. As noted earlier, the most useful method involves chromatography in detergents or nonionic chaotropes, although subunits may also be purified by sodium dodecyl sulfate polyacrylamide gel electrophoresis. Proteins extracted from polyacrylamide gels may contain impurities that interfere with amino acid analysis, particularly when the more sensitive methods are used.

Amino Acid Analysis

An amino acid analysis is the standard against which accumulated data from a sequence analysis is measured. Several determinations should be completed to insure an accurate composition. Generally proteins are hydrolyzed at 110°C in constant boiling HCl (6 *N*) in a sealed glass tube from which air has been removed by repeated evacuation after freezing the sample. Because of the skill required to seal these samples and the risk of loss of sample during freezing and thawing, Teflon lined screw capped tubes may be preferred. Phillips (167) describes an apparatus in which 12 protein samples can be prepared for hydrolysis in screw capped tubes in about 1 hr.

One of the problems encountered during the hydrolysis of proteins with HCl is the loss of tryptophan making it difficult to determine the number of tryptophanyl residues in a protein. Two different methods may be used to overcome this problem. In the first procedure, the protein is hydrolyzed under conditions that minimize loss of tryptophan such as 3 *N* mercaptoethane sulfonic acid, 3 *N* *p*-toluene sulfonic acid, 6 *N* HCl containing 2% thioglycolic acid (168), or 4 *N* methanesulfonic acid containing 0.2% (w/v) 3-(2-aminoethyl)indole (169). In the second procedure, the protein is first treated with performic acid to oxidize cysteine to cysteic acid and then hydrolyzed with 6 *N* HCl. Tryptophan is determined with a second sample using absorbance (170) or fluorescence measurements in 6 *M* guanidine·HCl (171), by second derivative spectroscopy (172), or by reaction with 3-diazonium-1,2,4,-(5-^{14}C)triazole (173). Tyrosine and cysteine may also be determined on the second sample. Tyrosine is determined by spectrophotometric methods (172) while cysteine is determined using reagents specific for sulfhydryl residues such as 5,5′-dithiobis(2-nitrobenzoic acid) (174), 6,6′-diselenobis(3-nitrobenzoic acid) (175), or 2,2′ or 4,4′ dithiopyridine (176).

Bohlen and Schroeder (162) prefer to hydrolyze two different protein samples. One in HCl containing 2% thioglycolic acid for analysis of all amino acids except cysteine and proline and the other, a performic acid oxidized sample, in 6 *N* HCl without thioglycolic acid to determine the remaining two amino acids. The loss of trytophan during hydrolysis in 6 *N* HCl with thioglycolic acid and of cysteine during performic acid oxidation was corrected for by analyzing the 98 amino acid peptide somatostatin as a standard.

Amino acids serine and threonine are estimated by extrapolating the values obtained at three hydrolysis times (24, 48, and 72 hrs) to zero time. Because peptide bonds flanked by valine and isoleucine hydrolyze slowly, their values are estimated from the 72-hr hydrolyzate.

The original automated methods for analyzing amino acids in a protein hydrolyzate involved separating the amino acids on an ion exchange column and then detecting them in the eluent after reaction with ninhydrin. In newer more sensitive procedures, the amino acids in the hydrolyzate are first derivatized and then separated for quantitation; dimethylaminoazobenzenesulfonyl chloride (160), *o*-phthalaldehyde (161), and phenylisothiocyanate (177) are used as derivatizing reagents.

Sequencing Strategy

Basically the strategy for determining the amino acid sequence of a protein is to first cleave it by two different methods to generate two sets of peptides, which are then isolated, purified, and sequenced. Comparing overlapping sequences provides the data necessary to determine the order of the peptides in the protein. The amino (178) and carboxyl terminal (179) amino acids of the protein identify the amino terminal and the carboxyl terminal peptides. An example of this strategy is presented in Reference (180).

This simple straightforward approach to sequencing is generally successful although experience with large proteins suggests that more exploratory experiments and planning lead to a more efficient determination. For example, Walsh et al. (158) found that most sequencing problems can be divided into several smaller, more solvable problems by exploiting the susceptibility of the original protein to limited proteolysis or to chemical cleavage at the rare asparginylglycine (Asn–Gly) and aspartylproline (Asp–Pro) peptide bonds. Their study of the sequence of phosphorylase *a*, a protein containing 841 residues, was greatly facilitated by an initial limited cleavage yielding two polypeptide chains.

Peptide Cleavage

Most investigators use proteinases to cleave proteins into their peptides. Chemical cleavages are less commonly used. Table 4.5 gives a list of the specificities of several useful proteinases and chemical cleavage reagents. Note that the cleavage reactions of some of the proteinases on this list

Table 4.5 Useful peptide cleavage methods

Cleaving Agents[a]	Reference	Specificity
Trypsin	156	Lysyl–X and arginyl–X bonds. If X is proline, cleavage is slow. Restrict cleavage to arginyl–X by first reacting protein with succinic, maleic, or citraconic anhydride. Modify arginyl–X residues with cyclohexan-1,2-dione in the presence of borate or with malonic dialdehyde (158) to restrict cleavage to lysyl–X.
α-Chymotrypsin	156	Tryptophanyl–X, tyrosyl–X, phenylalanyl–X, leucyl–X, and methionyl–X bonds. Cleavage is blocked if X is proline.
Staphylococcal proteinase	156	Glutamyl–X bonds. Glytamyl–proline bonds are not cleaved.
Cyanogen bromide	156, 181	Methionyl–X bonds
(2-(2-nitrophenyl-sulfenyl)-3-methyl-3'bromoindolenine) (BNPS-skatole)	156, 181	Tryptophanyl–X bonds
Dimethyl sulfoxide–HBr	181	Tryptophanyl–X bonds
o-Iodosobenzoic acid	156, 181	Tryptophanyl–X bonds
pH 2.5	158, 181	Aspartyl–proline bonds
Hydroxylamine	158, 181	Asparaginyl–glycine bonds

[a] The proteinases pepsin, thermolysin, subtilisin, and papain are too nonspecific to warrant general use. They may be used to cleave peptides when other approaches fail.

are not always complete; some peptide bonds flanked by amino acid residues of the correct specificity cannot be cleaved. Trypsin, for example, will not cleave lysyl– and arginyl–proline bonds; lysyl–lysine bonds are also partially resistant to cleavage. Cleavage patterns can also be changed by modifying the amino acid residues. Again, using trypsin as an example, the ϵ-amino groups of lysyl residues can be modified so that trypsin only cleaves arginyl peptide bonds. Citraconic anhydride is one of the most popular reagents for modifying the ϵ-amino groups of lysine. After cleaving the modified protein with trypsin and purifying the tryptic peptides, the blocking groups can be removed by incubating the peptide in dilute acetic acid overnight at room temperature (157).

Chemical methods of cleaving polypeptide chains are not as useful as enzymic methods because they lack the specificity of enzymes; the reactions often do not go to completion; and peptide side chains may be altered—glutamine and asparagine, for example, may be deamidinated. However, as the mechanisms of these reactions become more clearly understood and reaction conditions improve, it is expected that chemical cleavages will be used more extensively (181).

Cyanogen bromide is the most widely used reagent for cleaving peptide bonds. It cleaves on the C-terminal side of methionine to yield peptides with homoserine lactone in the C-terminal position. As with the proteolytic methods, certain peptide bonds are less susceptible to cleavage. For instance, methionyl–threonine and to a lesser extent, methionyl–serine bonds are often cleaved in low yield and methionyl–cysteine and *N*-acetylmethionyl peptides may not be cleaved at all.

The oxidative and brominating reagent BNPS-skatole is relatively specific for cleaving peptide bonds at tryptophanyl residues. Adding free exogenous tyrosine to reduce oxidation of tyrosine and histidine residues enhances the specificity of the cleavage reaction. *o*-Iodosobenzoic acid also effectively cleaves tryptophanyl–X bonds, but only when contaminating *o*-iodoxybenzoic acid is removed by incubation with *p*-cresol. The enhanced cleavage rate of aspartyl–proline bonds at pH 2 to 2.5 as compared to the other peptide bonds is due to the greater basicity of the proline nitrogen. Finally, under certain conditions, a cyclic imide (anhydro–aspartyl–glycyl) is formed at asparaginyl–glycine sequences, which is cleaved by hydroxylamaine to liberate a new amino-terminal glycine residue.

After the chemical or proteolytic cleavage is complete, centrifuge the enzyme digest to remove insoluble peptides known as core peptides. Eight molar urea or 10% formic acid will usually solubilize core peptides, which can then be fractionated by gel permeation chromatography into roughly 10 fractions. Peptides within each fraction are purified by ion exchange chromatography on either diethylaminoithyl (DEAE)-cellulose (for large peptides) or on Dowex type resins (for small peptides). References (156) and (157) give criteria and methods to estimate the purity of a peptide.

Special precautions are necessary when working with pure peptides. Cleanliness is essential; gloves should be worn to prevent contamination of the sample with amino acids and peptides in fingerprints. Autoclave all glassware to denature contaminating proteases and destroy microbial contamination. Store peptides at −20°C, preferably in the dry state.

Amino Acid Sequence

The Edman degradation is still the principal method used in sequencing peptides (157). It is a three step cyclic procedure carried out manually (182) or in an automatic instrument (156–158, 165, 166). In the first step, the amino terminal residue reacts with phenylisothiocyanate; in the second step, the reaction product is cleaved from the peptide via a cyclization reaction; and finally, the product, a thiazolinone derivative, is converted to the more stable phenylthiohydantoin and identified.

The Edman method is used in several different modes (158). In the subtractive mode, the composition of the shortened peptide establishes the missing amino terminal residue; in the dansyl mode, the new amino terminal residue is identified by end group analysis; and in the direct mode, the phenylthiohydantoin of the released amino acid is identified directly by various chromatographic techniques or as the free amino acid after hydrolysis. The automatic procedures use the direct mode (164).

As indicated above, identifying each amino acid in the sequence involves several steps. Consequently, it is important that each reaction go to completion because incomplete derivatization or hydrolysis of the derivative leads to release of different amino acids at each cycle of the process. Eventually several amino acids will be released at the end of each cycle and their relative stoichiometries will make it impossible to identify the correct amino acid. Yet it is important to point out that automatic sequenators are available that are extremely successful. Using the automatic direct mode, it is possible, starting with 10 nmol of protein, to sequence 70–90 residues; with 10 pmol, 15–25 residues can be sequenced (164). Because sequenators are expensive, resource centers at various universities specialize in sequencing peptides for a small fee.

Pinpointing modified amino acids in a sequence may require additional steps. For instance, cysteine residues that link different elements of the sequence through a disulfide bond (156, 157) and asparagine and glutamine (183) must be identified in the sequence. Electrophoretic methods are commonly used to locate these amino acids.

POSTTRANSLATIONAL COVALENT MODIFICATIONS

Contrary to the impression given by many biochemistry texts, proteins are not simple polymers of 20 amino acids. Uy and Wold (184) estimate

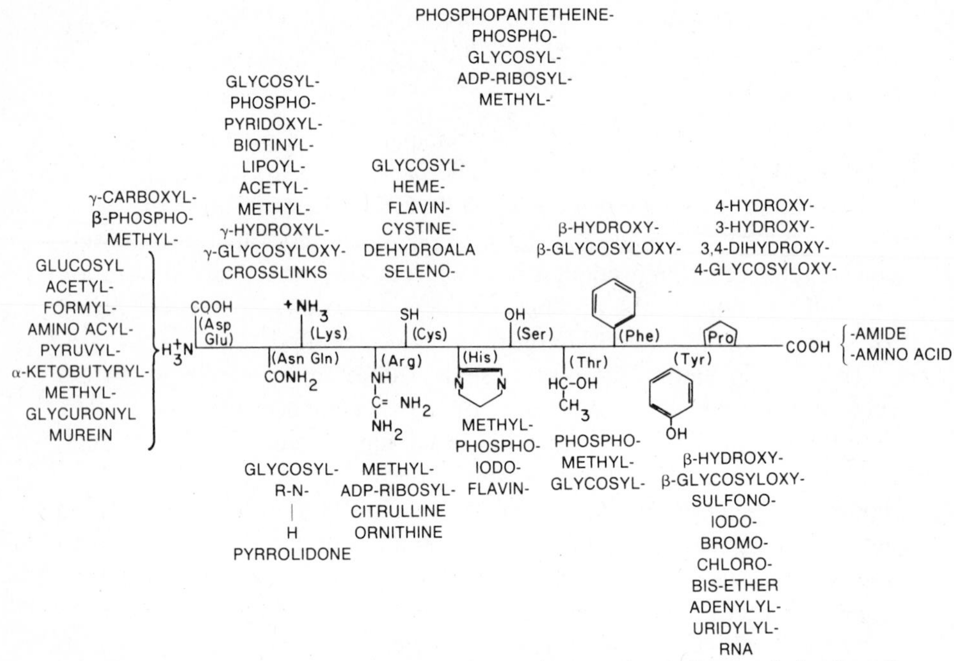

Fig. 4.17. Posttranslational modifications of proteins. Reprinted with permission from R. Uy and F. Wold, Posttranslational chemical modifications of proteins, in J. R. Whitaker and M. Fujimaki, Eds., *Chemical Deterioration of Proteins* (ACS Symposium Series 123), American Chemical Society, Washington, D.C., 1980, pp. 49–62. Reference (187).

that if all the proteins in an organism were hydrolyzed, several hundred different amino acid derivatives would be released. Modifications in vivo occur during polymerization on the polysomes (cotranslational), and after release of the protein precursor from the polysome (posttranslational) (185). For this discussion, all modifications will be referred to as posttranslational. Although derivatives of all the amino acids except alanine, glycine, isoleucine, leucine, methionine, valine, and tryptophan have been isolated (186), they will not be listed here. The different types of covalent modifications are summarized in Figure 4.17. For a more complete list, consult References (184, 187).

Because the universal initiation codon is AUG, all protein sequences should start with methionine or with *N*-formylmethionine depending on whether the protein is isolated from eucaryotes or procaryotes. Yet the N-terminal amino acid of most proteins is not methionine or *N*-formylmethionine. It is assumed, therefore, that the majority of all cellular proteins undergo a posttranslational proteolysis during or after synthesis. Other proteins, such as zymogens (proenzymes), prohormones, and proteins with leader peptides that initiate their transport across membrane barriers also undergo a posttranslational proteolysis.

In most posttranslational modifications, the amino acids in the polypeptide chain are covalently modified. The glycosyl derivatives of the primary amino acids, asparagine, serine, threonine, cysteine, as well as the secondary amino acids hydroxylysine, hydroxyproline, α-hydroxyphenylalanine, and α-hydroxytyrosine, form the largest group of posttranslational modified amino acids, although phosphorylated derivatives of serine, threonine, tyrosine, lysine, histidine, and arginine and the phosphate diesters linking guanyl and adenyl groups to serine and tyrosine are also common. The point to be made is that proteins are subjected to many different modifications during biosynthesis, and as a consequence, may contain many different amino acid derivatives. Identifying these amino acids in a protein before sequencing is important.

PEPTIDE MAPS

A peptide map is a one- or two-dimensional array of peptides obtained from an enzymic digest of a protein. Classically, peptide maps were two-dimensional arrays of peptides separated by electrophoresis on paper in one direction and then by solvent chromatography at right angles to the electrophoretic separation. Today peptide maps are prepared using several different techniques or a combination of techniques involving separation of peptides by high pressure liquid ion exchange and reverse phase chromatography, polyacrylamide gel electrophoresis, polyacrylamide or agarose gel isoelectric focusing, as well as other methods (188). A rapid and convenient method, often called the Cleveland technique, is especially suitable for proteins isolated from sodium dodecyl sulfate gels (189). It involves partial enzymic proteolysis in the presence of sodium dodecyl sulfate and analysis of the cleavage products by polyacrylamide gel electrophoresis. On a microscale, several hundred spots from 300 ng of protein can be resolved on two-dimensional peptide maps the size of a postage stamp (190). A more detailed discussion of high resolution peptide mapping may be found in Reference (191).

Peptide maps provide data for the solution of a variety of problems. In comparative enzymology, peptide maps are used as a sensitive indicator of differences in the amino acid sequence of a protein isolated from normal and diseased tissue. Pinpointing the difference between normal and sickle-cell hemoglobin (192) is a classical example of this methodology. Detecting and/or locating differences between subforms of enzymes caused by the deamidination of asparagines (193) or by chemical modification studies (194) is also common. Peptide maps are particularly useful for gaining information about the identity of the subunits of a protein. If a protein is composed of a single polypeptide chain, the number of spots on a peptide map is equal to the number of amino acid residues where cleavage is expected plus one. For a dimeric protein of identical subunits, the theoretical number is one half of the expected cleavages plus one. If the two subunits have different amino acid sequences, then the number of spots on the peptide map is equal to the number of amino acid residues where cleavage is expected plus two.

Cleaving Proteins

See the discussion under the heading Amino Acid Sequencing for details.

Detecting Peptides

Because the various peptide staining methods have different advantages and disadvantages, it is worthwhile to study them carefully before proceeding (188, 195). Either stain all peptides with a nonspecific stain or stain with reagents that are specific for individual amino acids. Nonspecific peptide stains are ninhydrin (188), fluorescamine (188), chlorination (196), and silver (189). Radioactive peptides can be detected with autoradiography (188, 197). Peptides containing specific amino acids can be detected either by modifying the protein before proteolysis or by staining the peptide array with specific stains. Modifying the protein with a chromogenic reagent before enzymic digestion provides a useful way of detecting the peptides with the naked eye (194) as the maps are developed.

Some specific amino acid stains used to stain peptides after preparation of the peptide map are the Pauly reaction for histidine and tyrosine (188), the Sakaguchi reaction (188), and the alkaline phenanthrenequinone (198) treatment for arginine, α-nitroso-β-naphthol stains tyrosine; platinic iodide brings out the peptides containing sulfur, and the Ehrlich reaction detects tryptophan (188).

TOPOGRAPHY OF A PROTEIN

A detailed description of the molecular surface of a protein, herein after called the topography of a protein, requires X-ray crystallographic data. Because such experiments require specialized equipment, suitable crystals, and a unique expertise, they will not be discussed in this book. Instead other methods and approaches for probing features of the surface of a protein molecule will be discussed.

Locating Functional Groups

Many amino acid residues in a protein have reactive functional groups. One of these residues, the sulfhydryl residue, is the subject of many investigations because it undergoes a facile disulfide interchange reaction. Determining the number of buried and exposed (surface) sulfhydryl residues in a native protein locates these functional groups thereby providing topographical information. For example, yeast pyruvate kinase contains five sulfhydryl residues per subunit; of this number, two are buried and not available for reaction in the native state, of the three exposed residues, one reacts rapidly, and the remaining two react more slowly but with the same first order rate constant (199). Adding the required cations, K^+ and Mg^{2+}, and the allosteric activator, fructose 1,6-diphosphate, changes the three-dimensional structure of the protein so that one of the slowly reacting exposed residues is no longer available for reaction.

A study of the sulfhydryl groups of muscle pyruvate kinase provides other topographical information. In the presence of its substrate, phosphoenolpyruvate, and the required metal ions, four sulfhydryl residues react with 5,5′-dithiobis(2-nitrobenzoic acid) (DTNB) to form disulfide links with 5-thio-2-nitrobenzoate. When the substrate and metal activators are removed, the thionitrobenzoate disulfide undergoes an intramolecular disulfide interchange as indicated in Figure 4.18. Two sulfhydryl groups per subunit of muscle pyruvate kinase are sufficiently close to one another to form a disulfide cross-link. The resulting pyruvate kinase is catalytically inactive (200).

In principle, it is possible to examine the location of all reactive protein functional groups; however, such topographical data in itself may not contribute significantly to the understanding of enzyme function. Occasionally, as with yeast pyruvate kinase (199), topographical data can be used to probe conformational changes in a protein. In addition, the oxi-

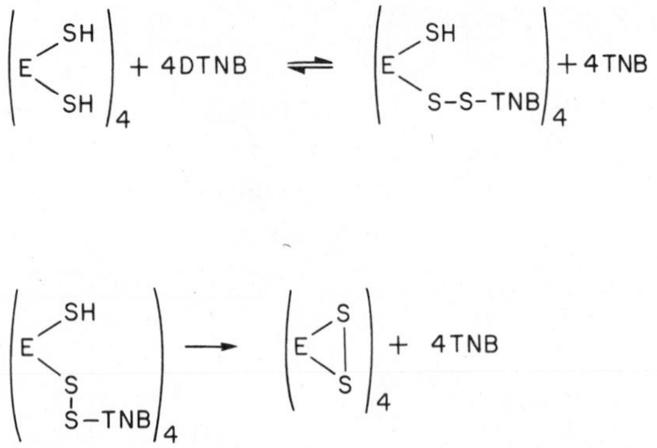

$$\left(E\begin{matrix}SH\\SH\end{matrix}\right)_4 + 4DTNB \rightleftharpoons \left(E\begin{matrix}SH\\S{-}S{-}TNB\end{matrix}\right)_4 + 4TNB$$

$$\left(E\begin{matrix}SH\\S\\|\\S{-}TNB\end{matrix}\right)_4 \longrightarrow \left(E\begin{matrix}S\\|\\S\end{matrix}\right)_4 + 4TNB$$

Fig. 4.18. Reaction of the sulfhydryl groups of muscle pyruvate kinase with 5,5′-dithiobis(2-nitro-benzoic acid) (200).

dation of adjacent sulfhydryl groups of muscle pyruvate kinase to form a disulfide may model physiologically significant thiol disulfide interchange reactions (201).

Functional groups in or near specific binding sites can be identified by affinity labeling techniques. Affinity labeling takes advantage of the affinity that a specific ligand has for its binding site, or in the case of enzymes, the active or allosteric site. Affinity labels are ligands that are modified by adding reactive groups or they are reactive reagents that mimic the geometry of the ligand binding site. When added to an enzyme, these affinity labels bind and react with functional groups in the vicinity of their binding site. Affinity labeling techniques include processes known as syncatalytic inactivation, active site directed inactivation, and mechanism based inactivation (202).

Already in 1963, affinity labeling reagents were used to pinpoint functional groups in the vicinity of the catalytic site of chymotrypsin. However, excitement about the use of these reagents diminished somewhat when a careful study of Staphylococcal nuclease using affinity labeling reagents showed a picture of the active site that was quite different from the picture determined by X-ray crystallography (203). The flexibility of the reagents or the protein appears to be responsible for many of the apparent differences in the active site topography. Data from affinity labeling studies, therefore, should be interpreted cautiously.

To increase the specificity of the affinity reagents, which is often sacrificed by the incorporation of a reactive functional group, use differential labeling methods (204). Differential labeling procedures require that the protein be reacted first with an affinity label in the presence of a tightly bound ligand (substrate, coenzyme, or competitive inhibitor) to mask its reaction with groups at the binding site. Removing the tightly bound ligand and then reacting the protein with the same affinity label in a radioactive form labels the site originally protected by the ligand.

Some investigators use photoaffinity labels to enhance specificity (205, 206). However, in addition to the labeling problems noted above, irradiated photoaffinity reagents, which have a long lifetime, label proteins nonspecifically, and in certain cases, label a number of different residues within a specific region of the binding site. This process is often called shotgun labeling. Irradiation itself may also inactivate the enzyme. In photoaffinity labeling studies then, a balance must be struck between the potential for nonspecific labeling and specific labeling; often maximum specific labeling must be sacrificed to limit nonspecific labeling.

Another type of affinity label is variously known as a suicide reagent, a mechanism based inactivator, or a k_{cat} inhibitor. Its specificity is determined not only by its binding affinity but also by its effectiveness in serving as a substrate for the target enzyme. The enzyme catalyzes the conversion of the substrate to a chemically reactive molecule, which then reacts with an active site moiety (207, 208). The efficiency at which a reagent inactivates an enzyme is determined by a partitioning ratio, that is, the average number of catalytic turnovers per inactivation event.

Locating Specific Peptides

When used in conjunction with electron microscopy, antibodies to components of the *E. coli* ribosome contribute to our understanding of the relative topography of the proteins in this complex (209, 210). Using proper controls, this methodology offers a unique approach to locating specific peptides on a surface. The recent development of methods for preparing monoclonal antibodies (211) makes the technique more powerful. Creating libraries of these antibodies, each of which recognizes a specific peptide, allows one to identify their topographical origin. Of course, monoclonal antibody methodology is limited if the peptides are nonimmune.

By judiciously selecting a proteinase or chemical reagent, it is possible, using proteolytic cleavage experiments, to visualize topographical features of a protein. For example, the first three enzymes of the pyrimidine biosynthetic pathway—carbamoyl phosphate synthetase, aspartate transcarbamylase, and dihydroorotase—copurify as a single subunit multifunctional protein during purification from several eucaryotes (212). Trypsin cleaves this multifunctional protein from hamster cells into three polypeptides, each responsible for one of the enzyme activities. This multifunctional protein, therefore, is composed of three independent polypeptide domains connected by a trypsin sensitive peptide sequence.

Proteolytic methods are also used to probe the topology of amphipathic membrane bound proteins. Trypsin, for example, cleaves membrane bound cytochrome b_5 reductase into a heme containing catalytically active hydrophilic peptide and a membrane bound hydrophobic peptide. The hydrophobic peptide contains 40 amino acids and is located on the carboxyl terminus (213). Brain hexokinase, the classic ambiquitous enzyme, binds to specific sites on the outer membrane of mitochondria; treatment with trypsin results in the release of a peptide from hexokinase and loss of binding (214). In a similar way, proteolytic methods are used to probe the actin recognition site on myosin (215) and the hemoglobin binding site on haptoglobin (216).

Locating Chromophoric Groups

Evidence for the topographical location of the chromophoric groups tyrosine and tryptophan in a protein can be obtained by spectroscopic methods. Changing the protein solvent perturbs the absorption spectrum of chromophoric groups on its surface; groups buried in the interior of the protein are shielded from the perturbing effect of the solvent (217). Determining the effect of quenchers I_2, O_2, acrylamide, Cs^+, or *N*-methylnicotinamide on the fluorescence of a protein also makes it possible to assess the degree of exposure of tryptophan to the solvent (218, 219). The bimolecular rate constant (k_q) for the quenching process provides a sensitive measure of the exposure of fluorescent tryptophanyl residues (219). If the tryptophanyl residue is sterically shielded from the quencher, this effect will be reflected in a decreased k_q as given by the Stern–Volmer equation

$$F_0/F = 1 + k_q T_0 \, (Q)$$

where F and F_0 are fluorescences in the presence and absence, respectively, of the quencher Q. The term $k_q T_0$, where T_0 is the fluorescence lifetime of the fluorophore in the absence of quencher, is the Stern–Volmer constant for the quenching reaction.

Proximity of Functional Groups

Reacting proteins with bifunctional reagents cross-link amino acid residues in juxtaposition to one another on the protein surface. The distance between the functional groups of the cross-linking reagent gives an estimation of the distance between the surface amino acid functional groups. Cross-linking experiments also aid in defining nearest neighbor relationships in multisubunit protein complexes and in "fixing" reversibly associating systems of proteins to obtain functional properties of the associated complex (220, 221).

Lactose synthase, for example, consists of two protein components that associate in a reversible manner to form 1:1 complexes during the catalytic cycle. α-Lactalbumin, the regulatory protein of the complex, promotes the binding of glucose to galactosyltransferase, the catalytic component of the complex. After reacting the complex with a homologous series of bifunctional bisimidoesters, functional complexes are obtained with the bifunctional reagents, dimethyladipimidate and dimethylpimelimidate; shorter bisimidoesters do not cross-link (222).

Recently, cross-linking reagents have been used to identify amino acid sequences involved in the actin–myosin complex. Cysteine 373, next to the C-terminal residue of actin, is first selectively labeled with a fluorescent dye. The labeled actin is then complexed with the tryptic subfragment 1 of myosin (S1), and subsequently cross-linked with a bifunctional reagent. After partially cleaving the cross-linked product and the control actin with cyanogen bromide and subjecting the products to electrophoresis, it is possible to locate the cross-linked site on the actin sequence vis á vis the C-terminal residue (223).

One large protein complex that is the subject of a variety of topographical studies is the ribosome (224), particularly the *E. coli* ribosome, which contains more than 50 proteins. To examine the nearest neighbor relationships among the protein subunits, a group of investigators recently reacted the ribosome with a bifunctional reagent containing a disulfide bond. The disulfide linked protein complexes were then fractionated by electrophoresis in polyacrylamide–urea gels. Following this fractionation, sequential slices of the urea gels were subjected to two-dimensional sodium dodecyl sulfate polyacrylamide gel electrophoresis; electrophoresis in the second direction was run in the presence of a reducing reagent to cleave the disulfide bond of the cross-linking reagent. Eleven interface cross-linked dimers, each containing one 50-S and one 30-S protein, were identified (225, 226).

Proximity of Other Groups

The transfer of fluorescence energy between a donor and acceptor molecule can be used as a spectroscopic ruler to measure distances between them. The energy transfer, which occurs over distances as long as 70 Å, depends on the inverse sixth power of the distance between the two molecules. The attractive features of this approach are its high sensitivity and its applicability to complex systems.

Stryer (38) has reviewed several applications of energy transfer experiments, among them studies showing that rhodopsin is an elongated molecule; that a conformational change in a receptor protein following the addition of a ligand is sensed by fluorescent groups 30-Å apart; and that distances between two metal ions in a protein can be measured. Applying this technique to the study of the 30-S *E. coli* ribosome, composed of 21 proteins, gives considerable insight into its structure.

Magnetic resonance techniques, including both electron paramagnetic resonance (EPR) and nuclear magnetic resonance (NMR), have emerged in the past two decades as powerful tools for mapping the surface of a protein and probing the mechanism of enzyme reactions. Many enzymes, for example, require the monovalent cation K^+ for catalysis. To determine whether K^+, required for pyruvate kinase catalysis, functions in the catalytic event per se or whether it is simply involved in maintaining an active conformation of the enzyme, Reuben and Kayne (227) examined the effect of the paramagnetic ion Mn^{2+} on the magnetic resonance signal of Tl^+, a useful K^+ substitute. Their data showed binding of Tl^+ within 4–8 Å of the required divalent cation at the active site of muscle pyruvate kinase. Magnetic resonance studies using other nuclei confirm these results (228, 229) clearly indicating that the monovalent cation provides an important function in the catalytic event.

It is possible to calculate distances up to 24 Å from a paramagnetic center from NMR data with a precision of better than ±10%, as determined by comparison with structures determined by crystallographic measurements (230). Distance measurements of similar magnitude can also be obtained by electron spin resonance (ESR) spectroscopy using spin labeled probes (231).

REFERENCES

1. M. J. Heeb and O. Gabriel, *Methods Enzymol.*, **104,** 416–439 (1984).
2. D. W. Kupke and T. E. Dorrier, *Methods Enzymol.*, **48,** 155–162 (1978).
3. T. L. Rosenberry, H. W. Chang, and Y. T. Chen, *J. Biol. Chem.*, **247,** 1555–1565 (1972).
4. B. Kickhofen and R. Warth, *Hoppe-Seyler's Z. Physiol. Chem.*, **357,** 745–749 (1976).
5. M. L. Bender, M. L. Begue-Canton, R. L. Blakeley, L. J. Brubacher, J. Feder, C. R. Gunter, F. J. Kezdy, J. V. Killheffer, Jr., T. H. Marshall, C. G. Miller, R. W. Roeske, and J. K. Stoops, *J. Am. Chem. Soc.*, **88,** 5890-5913 (1966).
6. D. V. Santi, C. S. McHenry, and E. R. Perriard, *Biochemistry,* **13,** 467–470 (1974).
7. J. W. Pierce, S. D. McCurry, R. M. Mulligan, and N. E. Tolbert, *Methods Enzymol.*, **89,** 47–55 (1982).
8. H. Theorell and T. Yonetani, *Biochem. Z.*, **338,** 537–553 (1963).
9. T. Yonetani and H. Theorell, *Arch. Biochem. Biophys.*, **99,** 433–446 (1962).
10. D. C. Livingston, J. R. Brocklehurst, J. F. Cannon, S. P. Leytus, J. A. Wehrly, S. W. Peltz, G. A. Peltz, and W. F. Mangel, *Biochemistry,* **20,** 4298–4306 (1981).
11. F. J. Kezdy and E. T. Kaiser, *Methods Enzymol.*, **19,** 3–20 (1970).
12. G. P. Lampson and A. A. Tytell, *Anal. Biochem.*, **11,** 374–377 (1965).
13. J. N. Behnke, S. M. Dagher, T. H. Massey, and W. C. Deal, Jr., *Anal. Biochem.*, **69,** 1–9 (1975).
14. R. M. Pino and T. K. Hart, *Anal. Biochem.*, **139,** 77–81 (1984).
15. T. Laas, I. Olsson, and L. Soderberg, *Anal. Biochem.*, **101,** 449–461 (1980).
16. P. G. Righetti, Isoelectric focusing: theory, methodology and applications, Vol. 11, *Laboratory Techniques in Biochemistry and Molecular Biology*, Elsevier, Amsterdam, 1983.
17. T. Laas and I. Olsson, *Anal. Biochem.*, **114,** 167–172 (1981).
18. K. W. Williams, T. Laas, and L. Soderberg, *Protides Biol. Fluids, Proc. Colloq.*, **27,** 677–681 (1979).
19. N. Y. Nguyen and A. Chrambach, *Anal. Biochem.*, **79,** 462–469 (1977).
20. N. Y. Nguyen and A. Chrambach, *Anal. Biochem.*, **82,** 226–235 (1977).
21. S. J. Cantrell, J. A. Babitch, and S. Torres, *Anal. Biochem.*, **116,** 168–173 (1981).
22. A. D. Friesen, J. C. Jamieson, and F. E. Ashton, *Anal. Biochem.*, **41,** 149–157 (1971).
23. W. W. Just, *Anal. Biochem.*, **102,** 134–144 (1980).
24. S. Binion and L. S. Rodkey, *Anal. Biochem.*, **112,** 362–366 (1981).
25. A. Pekkala-Flagan and D. E. Comings, *Anal. Biochem.*, **122,** 295–297 (1982).
26. D. Malamud and J. W. Drysdale, *Anal. Biochem.*, **86,** 620–647 (1978).
27. H. Edelhoch and R. F. Chen, The structural analysis of polypeptide and protein hormones by absorption and fluorescence spectroscopy, in C. H. Li, Ed., *Hormonal Proteins and Peptides*, Vol. 9, Academic, New York, 1980, pp. 109–173.

28. W. B. Gratzer, Optical methods for studying protein conformation, in *Techniques in Protein and Enzyme Biochemistry,* B108, Elsevier–North Holland, Amsterdam, 1978, pp. 1–43.
29. K. Rosenheck and P. Doty, *Proc. Nat. Acad. Sci. USA,* **47,** 1775–1785 (1961).
30. D. B. Wetlaufer, *Adv. Protein Chem.,* **17,** 303–390 (1962).
31. J. W. Donovan, Ultraviolet absorption, in S. J. Leach, Ed., *Part A, Physical Principles and Techniques of Protein Chemistry,* Academic, New York, 1969, pp. 101–170.
32. D. Freifelder, *Physical Biochemistry,* 2nd Ed., Freeman, San Francisco, 1982, pp. 573–593.
33. N. Greenfield and G. D. Fasman, *Biochemistry,* **8,** 4108–4116 (1969).
34. J. P. Hennessey, Jr., and W. C. Johnson, Jr., *Biochemistry,* **20,** 1085–1094 (1981).
35. P. Y. Chou and G. D. Fasman, *J. Mol. Biol.,* **115,** 135–175 (1977).
36. R. C. Hider and S. J. Hodges, *Biochem. Educ.,* **12,** 9–18 (1984).
37. W. Kabsch and C. Sander, *FEBS Lett.,* **155,** 179–182 (1983).
38. L. B. Stryer, *Annu. Rev. Biochem.,* **47,** 819–846 (1978).
39. F. W. J. Teale and G. Weber, *Biochem. J.,* **65,** 476–482 (1957).
40. R. F. Chen, *Anal. Biochem.,* **25,** 412–416 (1968).
41. R. P. Tengerdy and C. Chang, *Anal. Biochem.,* **16,** 377–383 (1966).
42. W. M. Vaughan and G. Weber, *Biochemistry,* **9,** 464–473 (1970).
43. G. Weber and L. B. Young, *J. Biol. Chem.,* **239,** 1415–1423 (1964).
44. D. C. Turner and L. Brand, *Biochemistry,* **7,** 3381–3390 (1968).
45. W. O. McClure and G. M. Edelman, *Biochemistry,* **5,** 1908–1919 (1966).
46. R. A. Kenner and A. A. Aboderin, *Biochemistry,* **10,** 4433–4440 (1971).
47. J. R. Barrio, J. A. Secrist III, and N. J. Leonard, *Biochem. Biophys. Res. Commun.,* **46,** 597–604 (1972).
48. N. J. Leonard, M. A. Sprecker, and A. G. Morrice, *J. Am. Chem. Soc.,* **98,** 3987–3994 (1976).
49. D. C. Ward, E. Reich, and L. Stryer, *J. Biol. Chem.,* **244,** 1228–1237 (1969).
50. L. A. Sklar, B. S. Hudson, and R. D. Simoni, *Proc. Nat. Acad. Sci. USA,* **72,** 1649–1653 (1975).
51. S. A. Latt, D. S. Auld, and B. L. Vallee, *Proc. Nat. Acad. Sci. USA,* **67,** 1383–1389 (1970).
52. Instructions to Authors, *J. Biol. Chem.,* **260,** 1–11 (1985).
53. R. J. Pollet, B. A. Haase, and M. L. Standaert, *J. Biol. Chem.,* **254,** 30–33 (1979).
54. W. W. Fish, *Methods Membrane Biol.,* **4,** 189–276 (1975).
55. H. Durchschlag and R. Jaenicke, *Biochem. Biophys. Res. Commun.,* **108,** 1074–1079 (1982).
56. G. L. Peterson and L. E. Hokin, *J. Biol. Chem.,* **256,** 3751–3761 (1981).
57. *CRC Handbook of Biochemistry,* 2nd Ed., H. A. Sober Ed., Chemical Rubber Co., Cleveland, OH, 1970, pp. C11–C23.
58. P. G. Martin, *Aust. J. Plant Physiol.,* **6,** 401–408 (1979).
59. R. E. Canfield, *J. Biol. Chem.,* **238,** 2698–2707 (1963).
60. M. O. Dayhoff, *Atlas of Protein Sequence and Structure,* Vol. 5, National Biomedical Research Foundation, Washington, D.C., 1972.

61. G. Frank and A. G. Weeds, *Eur. J. Biochem.*, **44,** 317–334 (1974).
62. T. Koide and T. Ikenaka, *Eur. J. Biochem.*, **32,** 417–431 (1973).
63. P. H. Corran and S. G. Waley, *Biochem. J.*, **145,** 335–344 (1975).
64. B. Andersson, P. O. Nyman, and L. Strid, *Biochem. Biophys. Res. Commun.*, **48,** 670–677 (1972).
65. D. Stone and L. B. Smillie, *J. Biol. Chem.*, **253,** 1137–1148 (1978).
66. C. Y. Lai, *Arch. Biochem. Biophys.*, **166,** 358–368 (1975).
67. J. H. Collins and M. Elzinga, *J. Biol. Chem.*, **250,** 5915–5920 (1975).
68. F. J. Castellino and R. Barker, *Biochemistry,* **7,** 2207–2217 (1968).
69. G. Zurawski, R. Perrot, W. Bottomley, and P. R. Whitfeld, *Nucleic Acid Res.*, **9,** 3251–3270 (1981).
70. K. G. Mann, W. W. Fish, A. C. Cox, and C. Tanford, *Biochemistry,* **9,** 1348–1354 (1970).
71. D. J. Graves and J. H. Wang, α-Glucan phosphorylases—chemical and physical basis of catalysis and regulation, in P. D. Boyer, Ed., *The Enzymes*, 3rd ed., Academic, New York, Vol. 7, 1972, pp. 435–482.
72. R. G. Martin and B. W. Ames, *J. Biol. Chem.*, **236,** 1372–1379 (1961).
73. D. S. Luthe, *Anal. Biochem.*, **135,** 230–232 (1983).
74. T. B. Nielsen and J. A. Reynolds, *Methods Enzymol.*, **47,** 3–10 (1978).
75. W. P. Campbell, C. W. Wrigley, and J. Margolis, *Anal. Biochem.*, **129,** 31–36 (1983).
76. J. F. Poduslo and D. Rodbard, *Anal. Biochem.*, **101,** 394–406 (1980).
76a. J. F. Poduslo, *Anal. Biochem.*, **114,** 131–139 (1981).
77. K. Lorentz, Anal. Biochem., **76,** 214–220 (1976).
78. L. Ornstein, *Ann. N.Y. Acad. Sci.*, **121,** 321–349 (1964).
79. B. J. Davis, *Ann. N.Y. Acad. Sci.*, **121,** 404–427 (1964).
80. B. D. Hames and D. Rickwood, Eds., *Gel Electrophoresis of Proteins: A Practical Approach*, IRL Press, Oxford, 1981.
81. H. R. Maurer, *Disc Electrophoresis and Related Techniques of Polyacrylamide Gel Electrophoresis*, 2nd ed., de Gruyter, Berlin, 1971, pp. 72–110.
82. A. Chrambach, T. M. Jovin, P. J. Svendsen, and D. Rodbard, Analytical and preparative polyacrylamide gel electrophoresis: An objectively defined fractionation route, apparatus and procedures, in N. Catsimpoolas, Ed., *Methods of Protein Separation*, Vol. 2, Plenum, New York, 1976, pp. 27–144.
83. J. S. Condeelis, *Anal. Biochem.*, **77,** 195–207 (1977).
84. J. C. H. Steele, Jr., and T. B. Nielsen, *Anal. Biochem.*, **84,** 218–224 (1978).
85. B. D. Hames, An introduction to polyacrylamide gel electrophoresis, in B. D. Hames and D. Rickwood, Eds., *Gel Electrophoresis of Proteins, A Practical Approach*, IRL Press, Oxford, 1981, pp. 1–91.
86. J. D. Kowit and J. Maloney, *Anal. Biochem.*, **123,** 86–93 (1982).
87. M. J. Leibowitz and R. W. Wang, *Anal. Biochem.*, **137,** 161–163 (1984).
88. K. von Hungen, R. C. Chin, and C. F. Baxter, *Anal. Biochem.*, **128,** 398–404 (1983).
89. Z. Wotjkowiak, R. C. Briggs, and L. S. Hnilica, *Anal. Biochem.*, **129,** 486–489 (1983).
90. J. M. Kittler, N. T. Meisler, D. Viceps-Madore, J. A. Cidlowski, and J. W. Thanassi, *Anal. Biochem.*, **137,** 210–216 (1984).
91. C. M. Wilson, *Anal. Biochem.*, **96,** 263–278 (1979).

92. R. F. Ritchie, J. G. Harter, and T. B. Bayles, *J. Lab. Clin. Med.*, **68,** 842–850 (1966).
93. R. E. Allen, K. C. Masak, and P. K. McAllister, *Anal. Biochem.*, **104,** 494–498 (1980).
94. M. Eschenbruch and R. R. Burk, *Anal. Biochem.*, **125,** 96–99 (1982).
95. S. Irie, M. Sezaki, and Y. Kato, *Anal. Biochem.*, **126,** 350–354 (1982).
96. M. Porro, S. Viti, G. Antoni, and M. Saletti, *Anal. Biochem.*, **127,** 316–321 (1982).
97. R. D. Friedman, *Anal. Biochem.*, **126,** 346–349 (1982).
98. L. D. Adams and D. W. Sammons, A unique silver staining procedure for color characterization of polypeptides, in R. C. Allen and P. Arnaud, Eds., *Electrophoresis '81, Advanced Methods—Biochemical and Clinical Applications*, de Gruyter, Berlin, 1981, pp. 155–166.
99. C. R. Merril, A silver stain workshop, in D. Stalhakos, Ed., *Electrophoresis '82*, de Gruyter, Berlin, 1983, pp. 826–842.
100. K. Ohsawa and N. Ebata, *Anal. Biochem.*, **135,** 409–415, (1983).
101. O. Gaal, G. A. Medgyesi, and L. Vereczkey, *Electrophoresis in the Separation of Biological Macromolecules*, Wiley, New York, 1980.
102. K. Weber and M. Osborn, Proteins and sodium dodecyl sulfate: Molecular weight determination on polyacrylamide gels and related procedures, in H. Neurath and R. L. Hill, *The Proteins*, Vol. 1, 3rd ed., Academic, New York, 1975, pp. 179–223.
103. K. Weber, J. R. Pringle, and M. Osborn, *Methods Enzymol.*, **26,** 3–27 (1972).
104. H. R. Maurer and R. C. Allen, *Z. Klin. Chem. Klin. Biochem.*, **10,** 220–225 (1972).
105. G. Palumbo and M. F. Tecce, *Anal. Biochem.*, **134,** 254–258 (1983).
106. G. Steck, P. Leuthard, and R. R. Burk, *Anal. Biochem.*, **107,** 21–24 (1980).
107. Y. W. Shing and A. Ruoho, *Anal. Biochem.*, **110,** 171–175 (1981).
108. B. F. Tack, J. Dean, D. Eilat, P. E. Lorenz, and A. N. Schechter, *J. Biol. Chem.*, **255,** 8842–8847 (1980).
109. R. M. Schultz and P. M. Wassarman, *Anal. Biochem.*, **77,** 25–32 (1977).
110. R. C. Pittman, S. R. Green, A. D. Attie, and D. Steinberg, *J. Biol. Chem.*, **254,** 6876–6879 (1979).
111. E. Alhanaty, M. Tauber-Finkelstein, and S. Shaltiel, *FEBS Lett.*, **125,** 151–154 (1981).
112. M.-C. Tzeng, *Anal. Biochem.*, **128,** 412–414 (1983).
113. G. Scheele, J. Pash, and W. Bieger, *Anal. Biochem.*, **112,** 304–313 (1981).
114. N. Hall and M. DeLuca, *Anal. Biochem.*, **76,** 561–567 (1976).
115. T. Wood and C. C. Muzariri, Anal. Biochem., **118,** 221–226 (1981).
116. O. Gabriel, *Methods Enzymol.*, **22,** 578–604 (1971).
117. B. D. Hames and D. Rickwood, Eds., Enzyme detection methods, in *Gel Electrophoresis of Proteins: A Practical Approach*, IRL Press, Oxford, 1981, pp. 254–263.
118. C. Broger, P. Allemann, and A. Azzi, *J. Applied Biochem.*, **1,** 455–459 (1979).
119. J. T. Whicher, J. Higginson, P. G. Riches, and S. Radford, *J. Clin. Pathol.*, **33,** 781–785 (1980).
120. H. A. Erlich, J. R. Levinson, S. N. Cohen, and H. O. McDevitt, *J. Biol. Chem.*, **254,** 12,240–12,247 (1979).
121. W. N. Burnette, *Anal. Biochem.*, **112,** 195–203 (1981).
122. T. K. Johnson, K. C. L. Yuen, R. E. Denell, and R. A. Consigli, *Anal. Biochem.*, **133,** 126–131 (1983).
123. R. Hawkes, *Anal. Biochem.*, **123,** 143–146 (1982).

124. B. Bowen, J. Steinberg, U. K. Laemmli, and H. Weintraub, *Nucleic Acids Res.*, **8,** 1–20 (1980).

125. G. Avigad, *Anal. Biochem.*, **86,** 443–449 (1978).

126. D. A. Knecht and R. L. Dimond, *Anal. Biochem.*, **136,** 180–184 (1984).

127. B. Tasheva and G. Dessev, *Anal. Biochem.*, **129,** 98–102 (1983).

128. D. Ochs, *Anal. Biochem.*, **135,** 470–474 (1983).

129. D. Rodbard and A. Chrambach, *Anal. Biochem.*, **40,** 95–134 (1971).

130. A Chrambach and D. Rodbard, Quantitative and preparative polyacrylamide gel electrophoresis, in B. D. Hames and D. Rickwood, Eds., *Gel Electrophoresis of Proteins: A Practical Approach*, IRL Press, Oxford, 1981, pp. 93–143.

131. D. Rodbard and A. Chrambach, *Proc. Nat. Acad. Sci. USA*, **65,** 970–977 (1970).

132. K. A. Ferguson, *Metabolism Clin. Exp.*, **13,** 985–1002 (1964).

133. J. L. Hedrick and A. J. Smith, *Arch. Biochem. Biophys.*, **126,** 155–164 (1968).

134. R. F. Doolittle, *Science*, **214,** 149–159 (1981).

135. J. A. Reynolds and C. Tanford, *J. Biol. Chem.*, **245,** 5161–5165 (1970).

136. H. Sandermann, Jr., and J. L. Strominger, *Proc. Nat. Acad. Sci. USA*, **68,** 2441–2443 (1971).

137. B. L. Anderson, R. W. Berry, and A. Telser, *Anal. Biochem.*, **132,** 365–375 (1983).

138. W. O. Richter, B. Jacob, and P. Schwandt, *Anal. Biochem.*, **133,** 288–291 (1983).

139. M. E. Himmel and P. G. Squire, *Int. J. Pept. Protein Res.*, **17,** 365–373 (1981).

140. M. Le Maire, E. Rivas, and J. V. Moller, *Anal. Biochem.*, **106,** 12–21 (1980).

141. C. H. Suelter, D. Thompson, G. Oakley, M. Pearce, H. D. Husic, and M. S. Brody, *Biochem. Med.*, **21,** 352–365 (1979).

142. W. W. Fish, K. G. Mann, and C. Tanford, *J. Biol. Chem.*, **244,** 4989–4994 (1969).

143. N. Ui, *J. Chromatogr.*, **215,** 289–294 (1981).

144. T. C. Laurent and J. Killander, *J. Chromatogr.*, **14,** 317–330 (1964).

145. J. Porath, *Pure Appl. Chem.*, **6,** 233–244 (1963).

146. G. K. Ackers, *Adv. Protein Chem.*, **24,** 343–446 (1970).

147. T. J. Mantle, Molecular weight determination by gel filtration density gradient, electrophoresis, and irradiation inactivation, in *Techniques in Protein and Enzyme Biochemistry*, B105, Elsevier–North Holland, Amsterdam, 1978, pp. 1–17.

148. J. O. Thomas, Determination of the subunit structure of proteins, in *Techniques in Protein and Enzyme Biochemistry*, B106, Elsevier–North Holland, Amsterdam, 1978, pp. 1–22.

149. J. K. Wang, E. E. Dekker, N. D. Lewinski, and H. C. Winter, *J. Biol. Chem.*, **256,** 1793–1800 (1981).

150. H. Sjostrom, O. Noren, L. Christiansen, H. Wacker, and G. Semenza, *J. Biol. Chem.*, **255,** 11,332–11,338 (1980).

151. B. Moss and E. N. Rosenblum, *J. Biol. Chem.*, **247,** 5194–5198 (1972).

152. F. K. Friedman, K. Alston, and A. N. Schechter, *Anal. Biochem.*, **117,** 103–107 (1981).

153. W. E. Brown and G. C. Howard, *Methods Enzymol.*, **91,** 36–41 (1983).

154. N. Ramachandran and R. F. Colman, *J. Biol. Chem.*, **255,** 8859–8864 (1980).

155. M. O. Dayhoff, W. C. Barker, and L. T. Hunt, *Methods Enzymol.*, **91,** 524–545 (1983).

156. G. Allen, Sequencing of proteins and peptides, in T. S. Work and R. H. Burdon, Eds., *Laboratory Techniques in Biochemistry and Molecular Biology*, Vol. 9, Elsevier–North Holland, Amsterdam, 1981.
157. L. R. Croft, *Introduction to Protein Sequence Analysis*, Wiley, Chichester, 1980.
158. K. A. Walsh, L. H. Ericsson, D. C. Parmelee, and K. Titani, *Annu. Rev. Biochem.*, **50,** 261–284 (1981).
159. C. H. W. Hirs, *Methods Enzymol.*, **91,** 3–8 (1983).
160. J-Y. Chang, R. Knecht, and D. G. Braun, *Methods Enzymol.*, **91,** 41–48 (1983).
161. P. Bohlen, *Methods Enzymol.*, **91,** 17–26 (1983).
162. P. Bohlen and R. Schroeder, *Anal. Biochem.*, **126,** 144–152 (1982).
163. A. S. Bhown and J. C. Bennett, *Methods Enzymol.*, **91,** 434–442 (1983).
164. A. S. Inglis, *Methods Enzymol.*, **91,** 443–450 (1983).
165. M. W. Hunkapiller, R. M. Hewick, W. J. Dreyer, and L. E. Hood, *Methods Enzymol.*, **91,** 399–413 (1983).
166. F. S. Esch, *Anal. Biochem.*, **136,** 39–47 (1984).
167. R. D. Phillips, *Anal. Biochem.*, **113,** 102–107 (1981).
168. B. Penke, R. Ferenczi, and K. Kovacs, *Anal. Biochem.*, **60,** 45–50 (1974).
169. R. J. Simpson, M. R. Neuberger, and T.-Y. Liu, *J. Biol. Chem.*, **251,** 1936–1940 (1976).
170. H. Edelhoch, *Hormonal Proteins and Peptides,* **9,** 109–173 (1980).
171. P. Pajot, *Eur. J. Biochem.*, **63,** 263–269 (1976).
172. J. C. Garcia-Borron, J. Escribano, M. Jimenez, and J. L. Iborra, *Anal. Biochem.*, **125,** 277–285 (1982).
173. M. C. DeTraglia, J. S. Brand, and A. M. Tometsko, *Anal. Biochem.*, **99,** 464–473 (1979).
174. P. W. Riddles, R. L. Blakeley, and B. Zerner, *Methods Enzymol.*, **91,** 49–60 (1983).
175. N. P. Luthra, R. B. Dunlap, and J. D. Odom, *Anal. Biochem.*, **117,** 94–102 (1981).
176. D. R. Grassetti and J. F. Murray, Jr., *Arch. Biochem. Biophys.*, **119,** 41–49 (1967).
177. R. L. Heinrikson and S. C. Meredith, *Anal. Biochem.*, **136,** 65–74 (1984).
178. J-Y. Chang, *Anal. Biochem.*, **102,** 384–392 (1980).
179. H. Tschesche, *Methods Enzymol.*, **47,** 73–84 (1977).
180. G. Allen and B. Wittman-Liebold, *Hoppe-Seyler's Z. Physiol. Chem.*, **359,** 1509–1525 (1978).
181. K-K. Han, C. Richard, and G. Biserte, *Int. J. Biochem.*, **15,** 875–884 (1983).
182. J-Y. Chang, *Methods Enzymol.*, **91,** 455–466 (1983).
183. R. E. Offord, *Methods Enzymol.*, **47,** 51–69 (1977).
184. R. Uy and F. Wold, *Science,* **198,** 890–895 (1977).
185. F. Wold, *Annu. Rev. Biochem.*, **50,** 783–814 (1981).
186. F. Wold, Posttranslational covalent modification of proteins, in M. Smulson and T. Sugimura, Eds., *Novel ADP–Ribosylations of Regulatory Enzymes and Proteins*, Elsevier–North Holland, Amsterdam, 1980, pp. 325–332.
187. R. Uy and F. Wold, Posttranslational chemical modification of proteins, in J. R. Whitaker and M. Fujimaki, Eds., *Chemical Deterioration of Proteins* (ACS Symposium Series 123), American Chemical Society, Washington, D.C., 1980, pp. 49–62.

188. G. T. James, *Methods Biochem. Anal.*, **26,** 165–200 (1980).

189. D. W. Cleveland, S. G. Fischer, M. W. Kirschner, and U. K. Laemmli, *J. Biol. Chem.*, **252,** 1102–1106 (1977).

190. R. O. Neukirchen, B. Schlosshauer, S. Baars, H. Jackle and U. Schwarz, *J. Biol. Chem.*, **257,** 15,229–15,234 (1982).

191. D. K. Aromatoria, J. Parker, and W. E. Brown, *Methods Enzymol.*, **91,** 384–391 (1983).

192. V. M. Ingram, *Biochim. Biophys. Acta,* **28,** 539–545 (1958).

193. B. Oray, M. Jahani, and R. W. Gracy, *Anal. Biochem.*, **125,** 131–138 (1982).

194. D. L. Brautigan, S. Ferguson-Miller, G. E. Tarr, and E. Margoliash, *J. Biol. Chem.*, **253,** 140–148 (1978).

195. N. G. Anderson and N. L. Anderson, *Anal. Biochem.*, **85,** 331–340 (1978).

196. C. W. Easley, *Biochim. Biophys. Acta,* **107,** 386–388 (1965).

197. K. H. Choo, R. G. H. Cotton, and D. M. Danks, *Anal. Biochem.*, **103,** 33–38 (1980).

198. S. Yamada and H. A. Itano, *Biochim. Biophys. Acta,* **130,** 538–540 (1966).

199. S-L. Yun and C. H. Suelter, *J. Biol. Chem.*, **254,** 1806–1810 (1979).

200. M. Flashner, P. F. Hollenberg, and M. J. Coon, *J. Biol. Chem.*, **247,** 8114–8121 (1972).

201. H. F. Gilbert, *J. Biol. Chem.*, **257,** 12,086–12,091 (1982).

202. B. V. Plapp, *Methods Enzymol.*, **87,** 469–499 (1982).

203. F. Wold, *Methods Enzymol.*, **46,** 3–14 (1977).

204. A. T. Phillips, *Methods Enzymol.*, **46,** 59–69 (1977).

205. V. Chowdhry and F. H. Westheimer, *Annu. Rev. Biochem.*, **48,** 293–325 (1979).

206. R. J. Guillory and S. J. Jeng, *Fed. Proc., Fed. Am. Soc. Exp. Biol.*, **42,** 2826 (1983).

207. R. R. Rando, *Methods Enzymol.*, **46,** 28–41 (1977).

208. C. Walsh, *Annu. Rev. Biochem.*, **47,** 881–931 (1978).

209. J. A. Lake and L. Kahan, *J. Mol. Biol.*, **99,** 631–644 (1975).

210. G. W. Tischendorf, H. Zeichhardt, and G. Stoffler, *Proc. Nat. Acad. Sci. USA,* **72,** 4820–4824 (1975).

211. P. D. Yurchenco, D. W. Speicher, J. S. Morrow, W. J. Knowles, and V. T. Marchesi, *J. Biol. Chem.*, **257,** 9102–9107 (1982).

212. J. N. Davidson, P. C. Rumsby, and J. Tamaren, *J. Biol. Chem.*, **256,** 5220–5225 (1981).

213. H. A. Dailey and P. Strittmatter, *J. Biol. Chem.*, **256,** 3951–3955 (1981).

214. J. E. Wilson, *Current Topics in Cellular Regulation,* **16,** 1–54 (1980).

215. D. Mornet, R. Bertrand, P. Pantel. E. Audemard, and R. Kassab, *Biochemistry,* **20,** 2110–2120 (1981).

216. J. W. Lustbader, J. P. Arcoleo, S. Birken, and J. Greer, *J. Biol. Chem.*, **258,** 1227–1234 (1983).

217. T. T. Herskovits, *Methods Enzymol.*, **11,** 748–776 (1967).

218. S. B. Omar and T. Schleich, Biochemistry, **20,** 6371–6378 (1981).

219. M. R. Eftink and C. A. Ghiron, *Anal. Biochem.*, **114,** 199–227 (1981).

220. K. Peters and F. M. Richards, *Annu. Rev. Biochem.*, **46,** 523–551 (1977).

221. J. W. DePierre and L. Ernster, *Annu. Rev. Biochem.*, **46,** 201–262 (1977).

222. S. K. Sinha and K. Brew, *J. Biol. Chem.*, **256,** 4193–4204 (1981).

223. K. Sutoh, *Biochemistry,* **21,** 3654–3661 (1982).

224. C. K. Kurland, *Annu. Rev. Biochem.*, **46,** 173–200 (1977).
225. J. A. Cover, J. M. Lambert, C. M. Norman, and R. R. Traut, *Biochemistry,* **20,** 2843–2852 (1981).
226. J. M. Lambert, R. Jue, and R. R. Traut, *Biochemistry,* **17,** 5406–5416 (1978).
227. J. Reuben and F. J. Kayne, *J. Biol. Chem.*, **246,** 6227–6234 (1971).
228. T. Nowak and C. H. Suelter, *Mol. Cell. Biochem.*, **35,** 65–75 (1981).
229. J. J. Villafranca and F. M. Raushel, *Fed. Proc., Fed. Am. Soc. Exp. Biol.*, **41,** 2961–2965 (1982).
230. M. Cohn and G. H. Reed, *Annu. Rev. Biochem.*, **51,** 365–394 (1982).
231. L. J. Berliner, *Methods Enzymol.*, **49,** 418–480 (1978).

5

Protein–Ligand Complexes and Enzyme Kinetics

All proteins exert their biological function through complexes formed with other compounds. Consequently enzymologists have expended considerable time and effort studying these complexes and the resultant effects of their formation. Therefore, one complete chapter is devoted to a discussion of them. For this discussion, compounds will be referred to as ligands, even though they may be substrates, other proteins, or cellular structures. For example, enzymes form complexes with their substrate(s), activators, or inhibitors. Hemoglobin binds O_2 and transports it from the lungs to tissues. Other proteins carry or transport their ligands across membrane barriers, and receptor proteins form complexes with hormones or other effectors to produce physiological responses.

Whether a complex forms or not depends on the concentration of its components and its dissociation constant K_D. Three types of complexes are defined for the purposes of this discussion: weak, strong, and very strong complexes. Weak complexes have K_D values larger than $10^{-3}M$; strong 10^{-3}–$10^{-7}M$; and very strong, $10^{-7}M$ and smaller. Because there is no clear distinction between the three different types of complexes, I shall not discuss them separately but rather focus on techniques that are used, first, to determine only the dissociation constants for a complex, second, to determine both the dissociation constant and the stoichiometry of a complex, and finally, to determine only the stoichiometry of a complex.

DETERMINING K_D ONLY

The methods that provide information for calculating the K_D but not the stoichiometry of a protein–ligand complex utilize kinetic, spectroscopic, affinity chromatographic, and affinity electrophoretic measurements. Under certain conditions, spectroscopic measurements can also be used to determine the stoichiometry of a complex. When choosing a method, one must consider the amount of protein and ligand available, their costs and properties, and the available instrumentation.

Kinetic Methods

Kinetic methods are often used to estimate the K_D for a protein–ligand complex. If the reversible first step of an enzymic catalyzed reaction is very fast compared to subsequent steps, the reaction obeys equilibrium kinetics. In contrast, steady state kinetics occurs when the rate of the first step is comparable to subsequent steps. If equilibrium kinetics apply to a single substrate enzyme catalyzed reaction, the K_m value may approximate the dissociation constant for the enzyme–substrate complex. However, as emphasized by Cornish-Bowden (1), assuming that equilibrium kinetics apply to an enzyme reaction without supporting evidence (2) is not a good practice. For our purposes, it will be assumed that the K_m is not equivalent to the thermodynamic dissociation constant. Yet it is important to know that some of the kinetic constants of multiple substrate reactions, and kinetic inhibition and activation constants are true dissociation constants. However, because the mechanisms of these reactions must be known a priori to evaluate their values and because space does not allow a more extensive discussion of these mechanisms, other texts must be consulted for more details (1, 3, 4).

Other kinetic methods for determining the K_D of a protein–ligand complex involve measuring the effect of increasing ligand concentrations on either the rate of reaction of a reagent with some protein functional group or on the rate of an enzymic reaction itself. The difficulty with this methodology, however, is that the reagent may react with more than one protein functional group and the ligand may affect the reactivity of the two groups differently. As a result, the pseudofirst order rate constants, plotted as a function of increasing ligand concentration, may show a more complex behavior than that expressed by the normal protein–ligand equilibrium (5).

Perturbation of a Protein Property

Newton's law states that for every action there is an equal and opposite reaction. Therefore, the interaction of a ligand with a protein to form a complex always perturbs the conformation and property of both the protein and the ligand. Whether or not this perturbation is observable depends on the magnitude of the conformative response and the sensitivity of the methods used to measure it.

The equations that describe the perturbation of a protein property as a function of ligand concentrations are derived as follows. Assuming the reaction

$$P + L \overset{K_D}{\rightleftharpoons} L_B$$

one can define a dissociation constant K_D as

$$K_D = (n[P_T] - [L_B])([L_T] - [L_B])/[L_B] \tag{5.1}$$

where n is the number of binding sites on the protein molecule, P_T is total protein, L_T is total ligand, and L_B the bound ligand; $(n[P_T] - [L_B])$ is the concentration of unoccupied binding sites while $([L_T] - [L_B])$ is the free ligand concentration. If it is now assumed that $[L_T]$ is much greater than $[P_T]$, Equation (5.1) becomes

$$K_D = (n[P_T] - [L_B])[L_T]/[L_B]$$

or, after rearranging,

$$[L_B]/n[P_T] = [L_T]/(K_D + [L_T]) \tag{5.2}$$

At this point it is useful to note the symmetry of these equations. For example, if $n[P_T]$ is interchanged for $[L_T]$ and $[L_T]$ for $[P_T]$ in Equation (5.2), the resulting Equation (5.3) has the same form.

$$[L_B]/[L_T] = n[P_T]/(K_D + n[P_T]) \tag{5.3}$$

One can, therefore, determine the K_D for a protein–ligand complex by either titrating a protein with ligand or a ligand with protein. Since $[L_T]$ must be much greater than $n[P_T]$ when titrating a protein with ligand, it follows that $n[P_T]$ must be much greater than $[L_T]$, when titrating a ligand with protein.

The reciprocal form of Equation (5.2),

$$n[P_T]/[L_B] = (K_D/[L_T]) + 1$$

shows that a plot of $n[P_T]/[L_B]$ versus $1/[L_T]$ gives a straight line with the slope equal to K_D. In perturbation methodology, however, $[L_B]$ and $n[P_T]$ are not measured directly. Rather it is assumed that the perturbation of a protein at a finite ligand concentration is proportional to $[L_B]$ and at infinite ligand concentration to $n[P_T]$. If the conformative response of a protein to the binding of a ligand is called δ, and the maximum response (at infinite ligand concentration) is called δ_{max}, it follows that

$$\frac{\delta}{\delta_{max}} = \frac{[L_B]}{n[P_T]}$$

Figure 5.1 shows a plot of $\Delta A_{295}/\Delta A_{295max}$ versus pM^{2+} for the interaction of Mg^{2+} and Mn^{2+} with pyruvate kinase. The change in absorbance of the enzyme brought about by the interaction of the divalent cation is ΔA_{295}; the maximum change in absorbance at an extrapolated infinite concentration of the cation is ΔA_{295max}. The midpoint of the titration curve gives the K_D for the pyruvate kinase–divalent cation complex. (The origin of the change in absorbance at 295 nm, and the advantages and disadvantages of treating data in this manner are discussed later in this chapter.)

Perturbation methodology is best suited for the study of weak and strong complexes. The K_D value can, in principle, be obtained by any measurement that reflects the perturbation of any property of the binding macromolecule. Measuring changes in absorbance (6, 7), fluorescence (8), or circular dichroism (9) of the protein are the most convenient. In each case, though, it is important to make certain that the maximum

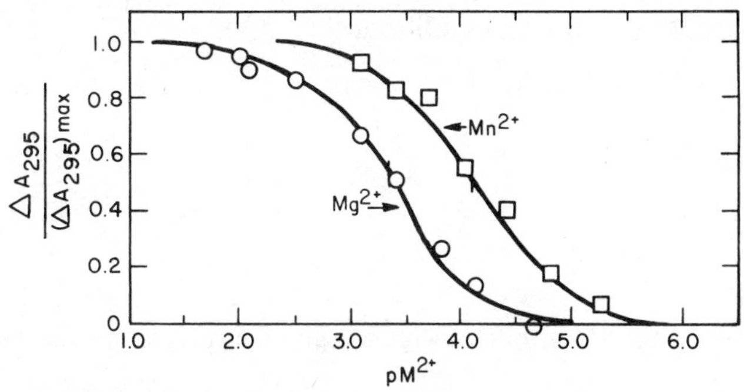

Fig. 5.1. The pM^{2+}-dependent difference spectrum of pyruvate kinase. Into each of two 3-mL fused silica cuvettes were placed 2 mL of an enzyme solution containing, per milliliter: Tris–HCl, pH 7.5, 50 μmol; KCl, 100 μmol; and pyruvate kinase in 5 m*M* TrisHCl, pH 8.2, 2.9 mg. Difference absorbances at 295 nm were read after addition of aliquots of either $MgCl_2$ or $MnCl_2$ to the sample cuvette and $(CH_3)_4NCl$ to the reference cuvette to maintain the same ionic strength. Abstracted by permission of the copyright owner, The American Society of Biological Chemists, Inc. Reference (6).

change in the measurement is directly proportional to protein concentration.

Monitoring the formation of strong complexes by perturbation methodology requires more sensitive techniques than those required to monitor weaker complexes. To ensure that $L = L_T$ at all points in an experiment, as required by this methodology, $n[P_T]/K_D$ must be small. However, the smaller the K_D, that is, the stronger the complex, the lower the protein concentration must be and the smaller will be the changes in the parameter used to measure the conformative response. Therefore, perturbation methodology is not applicable to very strong complexes.

Affinity Chromatography and Electrophoresis

Characterizing the interaction of a ligand with a protein by affinity chromatography, requires a chromatographic matrix with an immobilized ligand or a suitable analog of it. A small volume of protein is then added to the column containing the matrix and eluted with a solution of the competing ligand. As long as nonspecific factors do not affect elution of the protein, the volume of ligand solution required to elute the protein will depend on the concentrations of both the eluting and immobilized ligands and the dissociation constants of the protein for both ligands.

Assuming the following equilibria,

$$\begin{array}{l} \mathrm{P} + \mathrm{I{-}L} \overset{K_{\bar{D}}}{\rightleftharpoons} \mathrm{PI{-}L} \\ + \\ \mathrm{L} \\ \downarrow\!\uparrow K_{\mathrm{D}} \\ \mathrm{L_B} \end{array}$$

where I—L is the immobilized ligand, gives expression (5.4) (10).

$$1/(V_e - V_t) = (K_{\bar{D}}/[\mathrm{I{-}L}](V_t - V_0)) + (K_{\bar{D}}[\mathrm{L}]/K_{\mathrm{D}}[\mathrm{I{-}L}](V_t - V_0)) \quad (5.4)$$

The various volumes in Equation (5.4) are defined in Figure 4.14. A series of elutions of the protein, with differing concentrations of a competing ligand, from an affinity matrix with a fixed concentration of an immobilized ligand, provide the data necessary to calculate both K_{D} and $K_{\bar{D}}$ (10–12). K_{D} is the dissociation constant for the protein–mobile-ligand complex while $K_{\bar{D}}$ is for the protein–immobile-ligand complex. Both constants are obtained from a plot of $1/(V_e - V_t)$ versus L; $K_{\bar{D}}$ is calculated from the ordinate intercept, while K_{D} is obtained from the ratio of the ordinate intercept to the slope.

The derivation and use of Equation (5.4) is based on the following assumptions: (a) the concentration of the protein added to the column is much lower than the K_{D}(i.e., an insignificant amount of both the mobile or immobile ligand is complexed with the protein); (b) there is little nonspecific adsorption of the protein to the column matrix; and (c) the association and dissociation rate constants for formation of the complex are large compared to its rate of elution from the column (10).

Affinity electrophoresis combines the principles of electrophoresis with affinity chromatography. Electrophoresis is completed in polyacrylamide gels (13), in agarose, or in mixtures of agarose and polyacrylamide (14–16) containing different amounts of a macromolecular derivative of the ligand. The macromolecular-ligand derivative is physically entrapped in the gel and cannot migrate during electrophoresis. By determining the effect of varying concentrations of the mobile and/or immobile ligand on the electrophoretic mobility of the protein, one can determine the dissociation constant for the interaction of both the mobile and the immobile ligand with the protein.

The dissociation constants are evaluated with

$$K_D = [L]m_r/(m_{r0} - m_r) \tag{5.5}$$

where m_r is the relative electrophoretic mobility of the protein in the presence of a mobile ligand and m_{r0} is its mobility in the absence of the ligand (13). The relative electrophoretic mobility is the ratio of the distance migrated by the protein to the distance migrated by the tracking dye. The data can be evaluated from a plot of $1/m_r$ versus [L].

The derivation of Equation (5.5) is based on the following assumptions: (a) the protein concentration in the migrating zone is much lower than the mobile or immobile ligand concentration; (b) the protein–mobile-ligand complex and the free protein have the same relative electrophoretic mobility; (c) the protein–immobile-ligand complex has zero electrophoretic mobility; and (d) the rate of formation of the complex is large compared to the electrophoretic migration rate.

DETERMINING BOTH STOICHIOMETRY AND THE DISSOCIATION CONSTANT

Determining both the stoichiometry and the dissociation constant for a protein–ligand complex during the same experiment, requires that a significant fraction of the ligand be complexed with the protein at each data point in an experiment. Achieving a significant fraction of ligand complexation at practical protein concentrations however, requires that the complex have K_D values ranging from 10^{-3} to 10^{-8} M. This problem will be discussed in more detail later.

The equations that describe the interaction of a ligand with a protein in terms of both the K_D and K_{St}* and the stoichiometry of a complex are developed for a protein with two equal and independent binding sites as depicted in Figure 5.2. Because this hypothetical protein contains two

* The terms site dissociation constants (K_{D1}, K_{D2}, etc.) and stoichiometric dissociation constants (K_{St1}, K_{St2}, etc.) are used instead of intrinsic and extrinsic or microscopic and macroscopic constants (17). In addition, please note that whereas most authors use the lower case k to denote site constants and the upper case to denote stoichiometric constants, I have chosen to use the upper case to denote both constants. The lower case k will be used to denote rate constants.

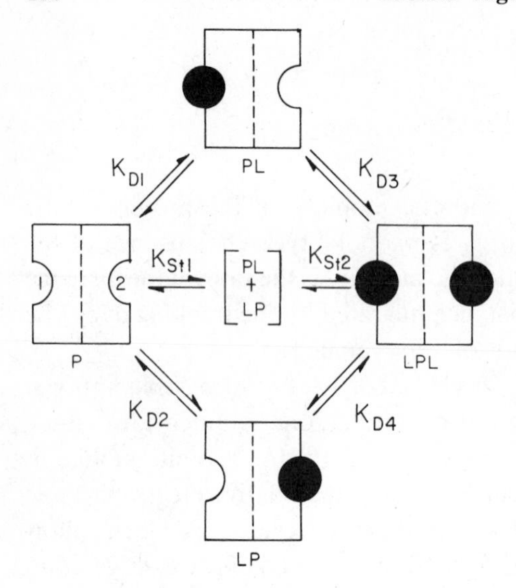

Fig. 5.2. Schematic diagram of a protein molecule with two ligand binding sites labeled site 1 and site 2. K_{D1}, K_{D2}, K_{D3}, and K_{D4} are site dissociation constants and K_{St1} and K_{St2} are stoichiometric dissociation constants. P symbolizes protein, LP, protein with site 2 occupied with ligand; PL, protein with site 1 occupied; and LPL, protein with both sites occupied with ligand.

sites, the ligand can bind at either site one or site two to form complexes PL or LP. The site constants for these two complexes are

$$K_{D1} = [P][L]/[PL] \qquad K_{D2} = [P][L]/[LP]$$

The first stoichiometric constant, as indicated in Figure 5.2 is given by

$$1/K_{St1} = ([PL]/[P][L]) + ([LP]/[P][L]) \tag{5.6}$$

or

$$1/K_{St1} = (1/K_{D1} + 1/K_{D2})$$

If K_{D1} and K_{D2} are identical, then $K_{St1} = K_{D/2}$.

The second stoichiometric constant K_{St2} equals

$$K_{St2} = ([PL][L]/[LPL]) + ([LP][L]/[LPL])$$

which when solved in terms of K_{D3} and K_{D4}, equals

$$K_{St2} = K_{D3} + K_{D4}$$

If the site dissociation constants K_{D3} and K_{D4} are also equal, then $K_{St2} = 2K_D$. When all four site dissociation constants are equal,

$$K_{St1}/K_{St2} = 1/4 \tag{5.7}$$

The general equations interrelating site and stoichiometric dissociation constants for the interaction of a ligand with a protein with n equal and independent binding sites are

$$K_{St1} = (1/n)K_D$$

$$K_{Sti} = ((i)/(n - i + 1))K_D$$

$$K_{Stn} = nK_D$$

The value of the ratio of the stoichiometric constants for the dissociation of protein multiligand complexes provides useful information about the mechanism of complex formation. The ratio 1/4 obtained from Equation (5.7) when all site constants have the same value reflects the statistical probabilities of the process. When K_{St1}/K_{St2} is greater than 1/4, the binding to the two sites is positively cooperative; when K_{St1}/K_{St2} is less than 1/4, the binding is negatively cooperative.

To determine the stoichiometry and the K_D for a protein–ligand complex, one must determine $\bar{v}$, defined by Equation (5.8), as a function of increasing ligand concentrations:

$$\bar{v} = \frac{\text{moles ligand bound}}{\text{moles protein}} = \frac{[L_B]}{[P_T]}$$

or (5.8)

$$\bar{v} = \frac{2[LPL] + [PL] + [LP]}{[P] + [PL] + [LP] + [LPL]}$$

Expressing each term in Equation (5.8) in terms of the site dissociation constants, when all site constants are equal, gives

$$\bar{v} = \frac{(2[P][L]^2/K_D^2 + 2[P][L]/K_D)}{[P] + 2[P][L]/K_D + [P][L]^2/K_D^2}$$

which can be simplified to give

$$\bar{\nu} = \frac{(2[\mathrm{L}]/K_{\mathrm{D}})(1 + [\mathrm{L}]/K_{\mathrm{D}})}{(1 + [\mathrm{L}]/K_{\mathrm{D}})^2} = \frac{2[\mathrm{L}]/K_{\mathrm{D}}}{1 + [\mathrm{L}]/K_{\mathrm{D}}}$$

or

$$\bar{\nu} = 2[\mathrm{L}]/(K_{\mathrm{D}} + [\mathrm{L}]) \tag{5.9}$$

For the general case with n sites (17)

$$\bar{\nu} = n[\mathrm{L}]/(K_{\mathrm{D}} + [\mathrm{L}]) \tag{5.10}$$

The application of this equation will be discussed later.

Equilibrium Dialysis

Equilibrium dialysis is the classical method for determining both the stoichiometry and the strength of a protein–ligand interaction. Klotz et al. (18) originally developed this method to study the interaction of organic ions with bovine serum albumin. Since then, equilibrium dialysis has been used extensively to study the interaction of ligands with proteins. To conduct a binding experiment by equilibrium dialysis, place the protein solution on one side of a dialysis membrane and the ligand solution on the other side in a closed apparatus as depicted in Figure 5.3 and allow the two solutions to equilibrate. After equilibrium is established, determine the concentration of ligand on both sides of the membrane. The moles of ligand bound $\mathrm{L_B}$ is given by

$$[\mathrm{L_B}] = ([\mathrm{L_p}] - [\mathrm{L_l}])V_{\mathrm{p}}$$

where $[\mathrm{L_p}]$ is the concentration of ligand on the protein side, V_{p} the volume of solution on the protein side, and $[\mathrm{L_l}]$ the concentration of ligand on the ligand side. Equation (5.8) is then used to calculate $\bar{\nu}$.

Although several equilibrium dialysis apparatus are commercially available, anyone with a minimum of machining experience can construct a unit as diagrammed in Figure 5.3 or described in the literature (19–22). The cell in Figure 5.3 contains 60 μL on a side. A dialysis membrane is placed between the two halves of the apparatus before joining them with

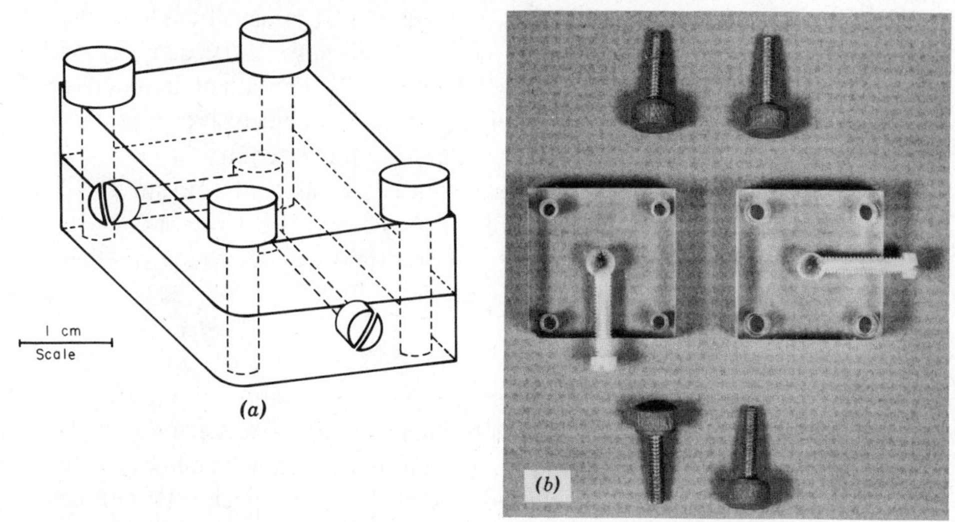

Fig. 5.3. (*a*) Drawing and (*b*) photograph of an equilibrium dialysis apparatus.

bolts. Wrapping the joined units with Teflon tape prevents leakage. If Visking dialysis membrane is used, it should be washed before use by boiling 5 min in 50 m*M* ethylenediaminetetraacetic acid (EDTA) and 5% $NaHCO_3$; it may also be stretched to increase porosity (21).

A successful binding experiment, using the apparatus given in Figure 5.3, requires careful attention to several details: (a) the dialysis chamber should not leak; (b) the protein should not cross the membrane barrier; (c) the protein should remain stable during the course of the experiment, that is, it should have the same biological activity at the start and conclusion of an experiment; and (d) the dialysis units must be rotated long enough to establish equilibrium. The ligand concentration on both sides of the membrane remains constant with time after equilibrium is established.

Two major problems plague equilibrium dialysis experiments. They are the time required for the concentration of highly charged molecules such as adenosine triphosphate (ATP) to come to equilibrium and Donnan membrane effects. The longer the time required for the ligand concentrations to come to equilibrium, the greater the loss of protein function, or if the ligand is not stable to hydrolysis, the greater its loss. The Donnan equilibrium is due to nondiffusable charged molecules, such as proteins, in one of the compartments (23). In practice, Donnan membrane effects may be diminished significantly by increasing the concentration of diffusible electrolytes in the binding reaction (0.1 *M* is usually sufficient). Alternatively, the effect of the Donnan equilibrium on the dialysis experiment can be determined by determining the distribution of radioactive cesium (24) and then correcting for its effect. If an electrolyte is added, make sure it does not affect the protein–ligand interaction.

Most equilibrium dialysis experiments are conducted with an equal volume of solution on both sides of the membrane. The protein concentration is measured before and after an experiment to determine whether the volumes of the compartments change.

Ultrafiltration

Ultrafiltration is the process in which small molecules are separated from macromolecules by filtration through a semipermeable membrane. In principle, it is equivalent to equilibrium dialysis. The concentration of the ligand that passes through the membrane is equal to its free concentration in the reaction mixture (25). Two ultrafiltration methods are in common use. In the first method (26), a solution containing protein and ligand in equilibrium is partially filtered; the concentration of ligand in the filtrate is the free ligand concentration. The moles of ligand bound is the difference between the sum of the moles of complexed and free ligand and the moles of free ligand.

In the second method, made popular by Paulus (27), the solution containing protein and ligand is forced through the semipermeable membrane by air pressure (28, 29) or by a swinging bucket ultracentrifugal filter (30). The ligand remaining on the membrane, minus a suitable blank, represents the moles of ligand bound; the concentration of ligand in the effluent is the free ligand concentration in the equilibrium.

Because the moles of ligand bound are determined directly in the Paulus procedure, rather than as the difference between moles of total and free ligand, the errors in $\bar{\nu}$ are reduced. Paulus (27) was able to determine both stoichiometry and K_D of a protein–ligand complex when less than 1% of the total ligand was bound. Although the sensitivity of the Paulus method is adversely affected by the amount of free ligand retained by the semipermeable membrane and the variation in this amount from sample to sample, it is possible to reduce these errors by regulating the transmembrane pressure during filtration and by controlling the time the filter remains subject to suction after all the liquid has passed through it (31). Careful selection of the membrane is also important (32).

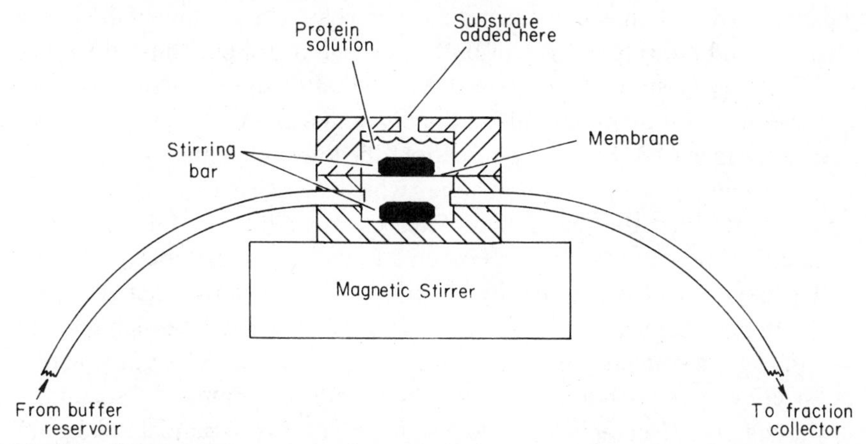

Fig. 5.4. Diagram of the apparatus for measuring substrate binding by rate of dialysis. The dialysis cell was adapted from the Technilab cell (Bel-Art Products) for continuous flow dialysis (1-mL size). The disc-shaped upper chamber (19 mm in diameter × 5 mm) was altered in two ways. It was deepened to 9 mm to make the capacity about 2.5 mL and a hole (5 mm in diameter) was drilled to the outside to permit additions of small volumes of substrate or other agents to the enzyme solution (1.5 mL) during the course of a binding measurement. The lower chamber (19 mm in diameter × 10 mm) has a capacity of 2.8 mL and is completely filled with buffer solution, which is pumped through at a constant rate of 8 mL/min. The membrane, a square cut from ordinary cellophane dialysis tubing (Union Carbide Company), is clamped between the Lucite blocks, which are held together by stainless steel screws. The contents of both chambers are mixed by means of small magnetic stirring bars; the bar in the top chamber rests on the membrane. The inner diameter of the tubing leading from the lower chamber to the fraction collector is small, 0.015 in., in order to minimize dead space. Aliquots (0.5 to 0.8 mL) from the fraction collector are diluted into a scintillator for liquid scintillation counting. The dialysis apparatus has been used satisfactorily either in the cold room (T = 5°C) or room temperature (T = 25°C). Abstracted by permission of the copyright owner, The American Society of Biological Chemists, Inc. Reference (33).

Rate of Dialysis

Because of the length of time required to complete an equilibrium dialysis experiment, Colowick and Womack (33) devised a binding experiment that could be completed in less than an hour. In this experiment, $\bar{\nu}$ is calculated from measurements of the rate of dialysis of a ligand across a semipermeable membrane. The dialysis rate is directly proportional to the free ligand concentration. The apparatus for measuring the rate of dialysis is shown in Figure 5.4. It consists of an upper chamber containing enzyme and radiolabeled substrate separated from the lower chamber by a semipermeable membrane. During an experiment, the contents of both chambers are stirred with a magnetic stirring bar. The volume of the lower chamber and the rate of flow of buffer through it can be adjusted so that different steady state levels of ligand in the effluent are possible.

Several useful modifications of the apparatus given in Figure 5.4 have been described (34). In one modification, the lower chamber was replaced with a small open faced spiral tube, which decreased the ratio of the volumes between the upper and lower chamber by about 10-fold (35), eliminating the need to stir the contents in the lower chamber.

The experiment is initiated by placing protein and a radioactive ligand in the upper chamber. The protein concentration should be high enough and the ligand concentration low enough so that at least 80% of the added radioactive ligand is bound. After sufficient buffer is pumped through the lower chamber to establish a steady state rate of loss of radioactive ligand from the upper chamber, a small volume of cold ligand is added to the upper chamber and a new steady state rate of loss of radioactive ligand from the upper chamber is determined. Successive additions of cold ligand are made, each followed by a measurement of the steady state rate of loss of radioactive ligand from the upper chamber, until less than 20% of the total radioactive ligand in the upper chamber is complexed with the protein. At this time, a large excess of cold ligand is added to the upper chamber and the final rate of dialysis of radioactive ligand is measured.

Figure 5.5 presents theoretical data for a typical rate of dialysis binding experiment. The steady state rate of loss of radioactive ligand from the upper chamber divided by the steady state rate after addition of excess cold ligand at the end of the experiment gives the fraction of the total ligand in the upper chamber that is free. This fraction, multiplied by the total amount of ligand added at each point in the experiment, gives the concentration of free ligand. In this way, sufficient data to calculate stoichiometry and K_D are obtained in less than an hour.

Selecting the most effective semipermeable membrane is the major problem associated with this method. The membrane must have the correct porosity so that the rate of dialysis is such that an insignificant amount of the ligand (preferably $<2\%$) is lost from the upper chamber during the experiment. Because of this restriction, the radioactive ligand must have a high specific activity. If less than 0.1% of the ligand dialyzes from the upper chamber per minute, 10^6–10^7 cpm is required in the upper chamber to obtain a significant number of counts in the lower chamber. Increasing the porosity of the membrane (36) decreases the time required to achieve steady state (34) and increases the sensitivity. Decreasing the volume of

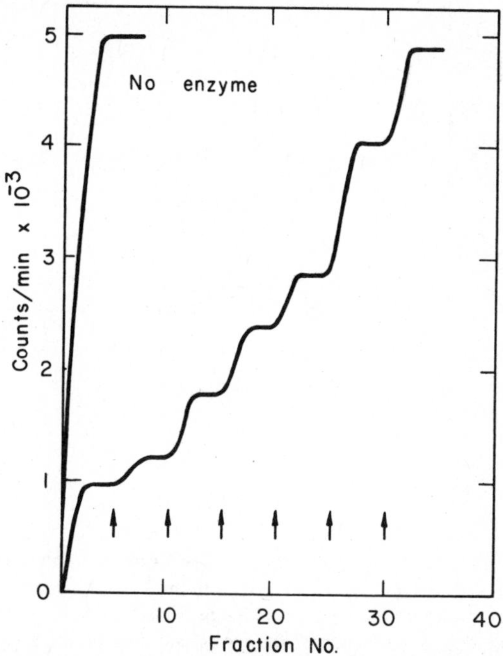

Fig. 5.5. Theoretical rate of dialysis curve for binding measurements. The ordinate shows calculated counts in effluent at each plateau after addition of unlabeled ligand. The counts are calculated assuming: $n[P_T] = 0.05$ mM; $K_D = 0.01$ mM; flow rate through lower chamber, 0.3 mL/min; volume upper chamber, 0.5 mL; volume lower chamber, 0.03 mL. Five samples are collected (1/min) after each addition of unlabeled ligand. At zero time (fraction zero), 5 nmol of ligand containing 10^7 counts is added. Cold ligand is then added to give final ligand concentrations in mM; 0.025, 0.05, 0.075, 0.1, 0.25, and 2.5. Assumed dialysis rate, 0.05% min^{-1}.

the lower chamber also increases the sensitivity, so that less radioactivity is required in the upper chamber. The increased sensitivity improves the economy of the experiment and allows for the determination of smaller dissociation constants.

The rates of dialysis must not be altered by factors, other than by changes in the concentration of free ligand, such as changes in the viscosity of the medium, changes in the permeability of the membrane, or changes in the rate of flow of buffer through the lower chamber (34). To guard against stirring related ionic exclusion effects, check whether the rate of stirring affects the steady state rate of dialysis (37). The ionic strength should also be maintained high enough to prevent Donnan membrane effects.

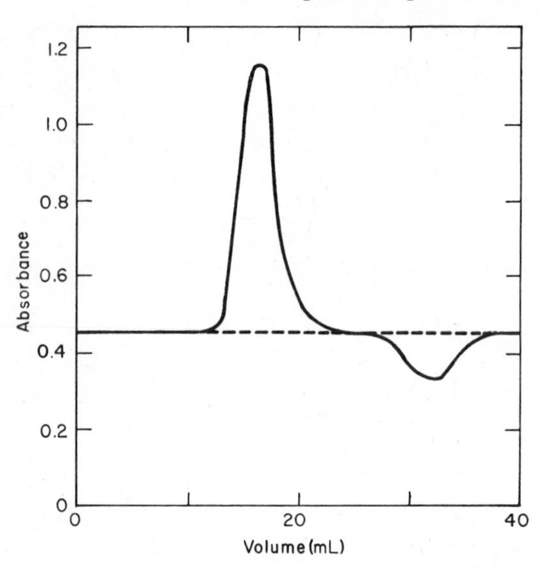

Fig. 5.6. Elution profile of pancreatic ribonuclease through a column of Sephadex G-25 equilibrated with 0.1 *M* acetate buffer, pH 5.3, containing cytidylic acid. Absorbance measurements at 285 nm (ordinate) indicate a positive peak (left) and negative trough (right) with respect to the (horizontal) baseline absorbance. Abstracted by permission from J. P. Hummel and W. J. Dreyer, *Biochim. Biophys. Acta*, **63,** 530–532 (1962) (39).

Gel Permeation Chromatography

Studying the binding of small molecules to proteins by gel permeation chromatography is particularly convenient (38). A typical binding experiment, as first described by Hummel and Dreyer (39), uses several gel permeation columns, each equilibrated with buffer containing a different concentration of ligand. A small protein sample is then prepared in each column equilibration buffer, added to the column, and then eluted with the same equilibration buffer. As indicated in the elution profile of Figure 5.6, the protein and its bound ligand are separated from the ligand depleted solvent during elution. It is essential that a plateau exist between the peak and trough area. If a plateau is not established, binding equilibria may be slow and the experiment should be repeated at lower flow rates (38). When a plateau cannot be established, the method should be discarded in favor of another.

The amount of ligand bound to the protein is equal to the excess ligand above the baseline (peak area) or the decreased ligand below the baseline (trough area). If possible, measure both the area of the trough and the peak region. In some situations, particularly those in which the ligand is measured by spectrophotometry and the protein and ligand both absorb light at the same wavelength, it is better to measure the area of the trough region. The free ligand concentration in the experiment is that used to equilibrate the column. The ratio of the moles of ligand equivalent to the trough area to the moles of protein applied to the column is $\bar{v}$.

Some investigators (40–42) use centrifugation to force the protein–ligand complex through a gel permeation column. Centrifuging a disposable 1-mL plastic tuberculin syringe containing the correct amount of Sephadex (determined by trial and error) separates the protein–ligand complex from the free ligand depleted solvent (trough area). The volume of solution forced out of the column contains the peak region while the free ligand depleted solvent (trough area) remains on the column. The data obtained by this column centrifugation technique may be more accurate than those obtained with the normal elution methods (42).

If the ligand absorbs light in the ultraviolet or visible region, its concentration in the trough and peak regions may also be determined by optically scanning the column before the peak is eluted, using a precision bore chromatographic column and special optical scanning equipment (38). In this Brumbaugh–Ackers method, the column is first equilibrated with a ligand and then scanned to determine its effective extinction coefficient. As before, a sample volume of a protein solution prepared with the column equilibration buffer is applied to the column, but instead of eluting the protein–ligand complex from the column, a sufficient volume of the column equilibration buffer is added to move the protein and protein–ligand complex through the major part of the column. Before the peak begins to elute from the column, the column flow is stopped and the column is scanned at the appropriate wavelength through its entire length to determine the concentration of ligand in both the trough and peak regions. Repeat this procedure with columns equilibrated with different concentrations of ligand.

Ultracentrifugation

Sedimenting protein from a solution with its ligand using a preparative ultracentrifuge (43) or airfuge gives a supernatant liquid containing ligand at its free concentration. The experiments may also be completed using an analytical ultracentrifuge employing a Tiselius partition cell (44). The solution containing ligand, but no protein, should be sedimented as a control. The major difficulty with this method is that the smaller the protein, the longer the time required to sediment it.

Perturbing a Ligand Property

If some property of a ligand is perturbed in a measurable way when bound to a protein, this perturbation can be used to determine both its stoichiometry and dissociation constant. Equation (5.11)

$$\delta = (x_p - x_0) \tag{5.11}$$

defines the change in the property of a bound ligand, given by the symbol δ, as the difference between the measured value of some property of a ligand in the presence of a protein (x_p) and that measured in the absence of protein (x_0) (45). The change in this property per mole of bound ligand, ϵ, is related to δ by

$$\delta = \Delta\epsilon[L_B]l \tag{5.12}$$

where $[L_B]$ is the concentration of bound ligand and l is the cuvette pathlength. The free ligand concentration [L] is obtained from the conservation equation

$$[L_T] = [L] + [L_B] \tag{5.13}$$

The usual procedure for evaluating $\Delta\epsilon$ is to titrate a fixed amount of ligand with protein. Extrapolating a plot of $1/\delta$ versus $1/[P_T]$ to infinite protein concentration gives δ for the bound ligand, from which $\Delta\epsilon$ is calculated using Equation (5.12). It is important to realize that the intercept of a double reciprocal plot of $1/\delta$ versus $1/[P_T]$ underestimates the desired value of δ unless $[L_T]/K_D$ is small (46); the maximum size that this ratio can be before the experiment becomes invalid will be discussed later. As noted earlier by Equations (5.2) and (5.3), the symmetry of the equilibrium equation allows one to determine the K_D of a weak complex either by titrating a protein with ligand when $(n[P_T]/K_D)$ is small or a ligand with protein when $[L_T]/K_D$ is small. Titrating a protein first with ligand and then a ligand with protein provides a useful way to confirm the K_D value for a protein–ligand complex (47).

Experiments in which spectrophotometric or fluorometric properties of a ligand are perturbed following its interaction with a protein are easy to set up and execute. As early as 1959, Stockell (48) utilized the change in absorbance of bound nicotinamide adenine dinucleotides (NAD^+) to determine both the stoichiometry and the K_D for its interaction with glyceraldehyde-3-phosphate dehydrogenase. The change in fluorescence of bound nicotinamide adenine dinucleotide (reduced) (NADH) can also be used to measure its interaction with a protein.

Some of the most exciting applications of ligand perturbation methodology involve the use of fluorescence probes. After a probe interacts with a protein, its fluorescence is either quenched or enhanced. 1-Anilino-8-napthalene sulfonate (ANS) and its derivatives form a classical group of probes whose fluorescence is enhanced when bound (see Chapter 4). Apomyoglobin binds 1 mol of the fluorescence probe per mole of protein with a near 200 fold increase in quantum yield. The probe binds at the heme binding site because hemin displaces it (49). Another probe, 1,N^6, etheno-cAMP, a highly fluorescent derivative of cyclic adenosine monophosphate (cAMP), is quenched 85% when bound to the regulatory subunit of type one cAMP dependent protein kinase. Since cAMP displaces the bound fluorescent probe, this method provides a convenient way to determine the stoichiometry of cAMP binding (50). The spectroscopic properties of some lanthanides make them useful as probes for the binding of biologically important divalent cations (51). For example, the lanthanide terbium displaces Ca^{2+} from its single binding site on porcine trypsin; the fluorescence of bound terbium is enhanced 10^5-fold.

Nuclear magnetic resonance spectrometry is particularly useful for detecting the perturbations of paramagnetic ligands. Paramagnetic ions possess electric quadrupole moments, which affect the longitudinal nuclear magnetic relaxation rate $1/\tau_1$ of the protons of water (52). The relaxation rate of the water protons is a weighted average of the relaxation rates of protons in different environments. Protons within the hydration sphere of a paramagnetic ion make the dominant contribution to the relaxation rate. When the paramagnetic ion is bound to a macromolecule, the relaxation rate of the bound water is enhanced (53, 54). By taking advantage of this enhancement, it is possible to obtain the necessary data to calculate both stoichiometry and K_D.

Perturbing the Protein

Determining the K_D for a protein–ligand complex by following the change in property of a protein as a function of increasing ligand concentration was outlined earlier in this chapter. This method was applied when in-

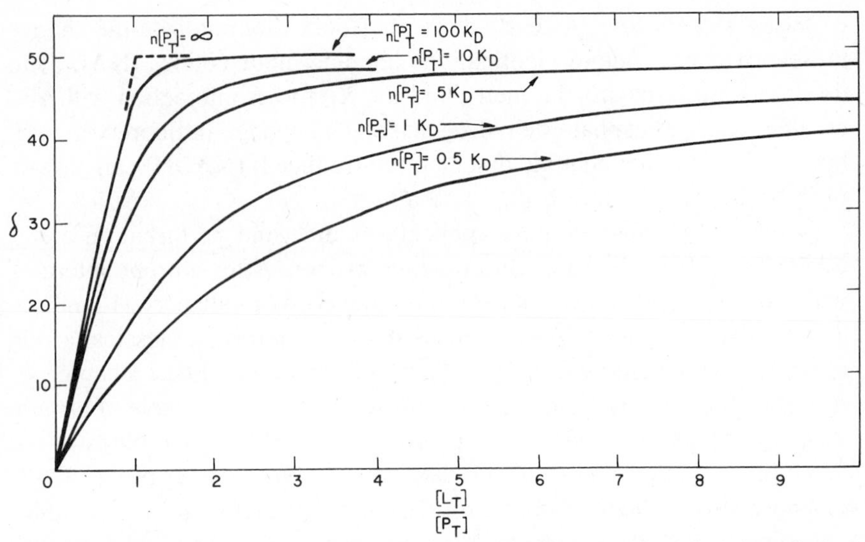

Fig. 5.7. Computed curves using Equation (5.2) of δ versus $[L_T]/[P_T]$ for a hypothetical protein with n ligand binding sites; $\delta_{max} = 50$.

significant amounts of ligand were bound during the titration (Figure 5.1). However, as will be discussed next, it is possible under other conditions to determine both the stoichiometry and the dissociation constant for a complex from protein perturbation measurements.

To understand how the perturbation of a protein molecule can be used to determine both stoichiometry and K_D, examine Figure 5.7 which shows that as $n[P_T]$ becomes larger, a plot of δ versus $[L_T]/n[P_T]$ approaches two straight lines which intersect at $[L_T]/n[P_T]$ equal to n, the stoichiometry of the interaction. Therefore, completing two experiments, one at high protein concentration and one at low protein concentration, provides data to determine both the stoichiometry and the dissociation constant for a complex. On the other hand, these same binding constants can also be calculated with data obtained from two experiments at intermediate concentrations of protein (Figure 5.7).

Dividing both sides of Equation (5.13) by $[P_T]$ and rearranging, yields

$$[L_B]/[P_T] = [L_T]/[P_T] - [L]/[P_T] \tag{5.14}$$

Solving for $[L]$, after equating two expressions of Equation (5.14) at two different protein concentrations $[P]_a$ and $[P]_b$, when $[L_B]/[P_T]$ values are identical (as indicated by identical δ values), gives

$$[L] = ([L_T]_b[P]_a - [L_T]_a[P]_b)/([P]_a - [P]_b) \tag{5.15}$$

where $[L_T]_a$ and $[L_T]_b$ are the total concentrations of ligand in the two experiments with protein concentrations $[P]_a$ and $[P]_b$, respectively. Knowing that

$$\bar{\nu} = [L_B]/[P_T] = [L_B]/[P_a] = [L_B]/[P_b]$$

and substituting $\bar{\nu}$ into Equation (5.14), and solving for [L] yields

$$[L] = [L_T]_a - \bar{\nu}[P]_a \tag{5.16}$$

Equating Equations (5.15) and (5.16) and rearranging yields Equation (5.17):

$$\bar{\nu} = ([L_T]_a - [L_T]_b)/([P]_a - [P]_b) \tag{5.17}$$

The values for $\bar{\nu}$ and [L] are then calculated at several points through the two titrations where the δ values are identical at the two protein concentrations $[P]_a$ and $[P]_b$ and analyzed as discussed later to give n and K_D. This approach has an advantage over a simple spectroscopic perturbation measurement because it does not require a measure of δ at infinite concentrations of the ligand (55–57).

Miscellaneous Methods

Other methods for studying protein–ligand interactions are available, but for various reasons, they are seldom used. For example, calorimetric data can be used to determine stoichiometry and dissociation constants for a protein–ligand interaction. These measurements, however, are not often made, presumably because they require specialized equipment (58, 59). Protein ligand interactions may also be studied by partitioning a binding reaction between two immiscible solvent phases. If the protein and protein–ligand complex are preferentially soluble in one phase while the ligand is distributed more or less evenly in both phases, measuring the distribution of the ligand yields information about the protein–ligand interaction (60, 61). Because of the difficulty of finding suitable solvent systems, this method fails to attract much use. Finally, methods using specific electrodes are some of the most practical for measuring the concentration of free ligand (62–64). Hydrogen ion electrodes are used daily to measure the concentration of free H^+, but the availability of other ion specific electrodes is limited.

Very Strong Complexes

Most of the methods for studying very strong complexes were developed by researchers interested in studying receptors for pharmacologic agents. Receptors are proteins that recognize specific ligands which, when bound, initiate a cellular response. Since these receptors do not exhibit enzymic activity, pharmacologists expended considerable time and effort to develop methods to determine their concentration and K_D values for the binding of ligands. As a result, most studies of very strong complexes are recorded in pharmacological literature.

The major difficulty in studying very strong complexes is nonspecific binding, that is, binding of a ligand that does not evoke a biological response. Two types of nonspecific binding are described. The first type, called strictly nonspecific, describes binding of a radiolabeled ligand to a site for which the parent ligand does not compete. The second is called specific nonreceptor binding. In this case the ligand binds in a specific manner to a nonreceptor macromolecule.

Specific and nonspecific binding are usually distinguished by studies of ligand affinities and biological responses (65, 66). When the ligand binding curve is compared with the biological response curve, the part of the curve that coincides with the biological response describes specific receptor binding. Ligand binding that does not evoke a biological response is nonspecific binding or specific nonreceptor binding. In an intact cell preparation that exhibits a biological response, it may be relatively easy to distinguish between these two types of binding. However, with an isolated preparation, which does not give a biological response, it is more difficult to distinguish between them. Techniques for overcoming these difficulties are discussed by Hollenberg and Cuatrecasas (65).

A common method for distinguishing between specific and nonspecific binding employs the principle of isotopic dilution. Measuring the amount of radioactive ligand bound to a receptor both in the presence and absence of excess unlabeled ligand gives a measure of nonspecific binding. The excess unlabeled ligand should effectively compete with the radiolabeled ligand for the receptor site. First add radiolabeled ligand to the receptor preparation and allow sufficient time for equilibrium to be established. Then add a 50-fold excess of unlabeled ligand to displace the labeled ligand from the receptor site. Radioactive ligand that is not displaced is bound in a strictly nonspecific manner.

A serious problem arises if the affinities of labeled and unlabeled ligand are different. For example, making the ligand radioactive may alter its binding property or the labeled ligand may be significantly contaminated with a degradation product that has a different affinity for the receptor. Sometimes the labeled ligand is not stable. By using a bioassay, however, it is possible, in principle, to determine whether there is a difference in the affinities of labeled and unlabeled ligand. The binding curve for radiolabeled ligand should coincide with the binding curve for unlabeled ligand monitored by the bioassay. For a variety of reasons though (primarily because of high nonspecific binding), it is usually not feasible to determine an entire binding curve with radiolabeled ligands. To overcome this difficulty Hollenberg and Cuatrecasas (65) suggest two basically different methods. In the first of these, a labeled ligand is added to a receptor preparation and the competitive effect of the unlabeled ligand on the binding of a fixed amount of labeled ligand is determined. The dissociation constant is calculated with

$$K_D = [IC]_{50}/(1 + ([L']/K'_D)) \qquad (5.18)$$

where $[IC]_{50}$ is the concentration of unlabeled ligand when 50% of the labeled ligand $[L']$ is displaced (65, 67). Note that an evaluation of the K_D for unlabeled ligand requires a knowledge of the K_D for labeled ligand K'_D. Furthermore, this equation only applies when the concentration of the receptor relative to the K_D is small (see Figure 5.3). If $[\text{receptor}]/K_D = 0.1$ or greater, then a significant fraction of the unlabeled ligand will be complexed during the titration and the observed $[IC]_{50}$ will be larger than the true $[IC]_{50}$.

To eliminate the need for determining the K_D of a labeled ligand, complete two (or preferably more) experiments on the competitive effect of increasing unlabeled ligand concentrations on the binding of different but fixed concentrations of labeled ligand. Plotting 1/(counts per min) of labeled ligand bound versus the ligand concentration (the Dixon plot) yields, at the abscissa value of the point where all extrapolated lines intersect, the value of the dissociation constant for the unlabeled ligand (65).

In the second method, called the dose ratio method, the binding of an increased concentration of a radioactive ligand is determined both in the presence and absence of a fixed concentration of unlabeled ligand. The ratio of the concentration of labeled ligand necessary to achieve a given amount of binding in the presence of unlabeled ligand $[X_1]$ to the concentration necessary to achieve the same amount of binding in the absence of unlabeled ligand $[X_2]$ is measured to give the dose ratio $DR = [X_1]/[X_2]$. The theory and graphical treatment of the dose ratio method are described in more detail by Chang et al. (68).

It is important to recall, at this point, the relationship between the rate constants for the formation and breakdown of a complex and the equilibrium constant for a reaction. It is known that for the equilibrium

$$P + L \underset{k_2}{\overset{k_1}{\rightleftharpoons}} PL$$

$K_D = k_2/k_1$ and the association rate constant k_1 is often diffusion controlled, tending to be relatively constant for all small molecule–macromolecule interactions. Assuming

$$k_1 = 10^7/\text{sec } M \qquad \text{and} \qquad K_D = 10^{-9}M,$$

$$k_2 = 10^{-2}/\text{sec or } t_{1/2} = 69.3 \text{ sec; for } K_D = 10^{-12}M,\ t_{1/2} = 19.3 \text{ hr.}$$

Therefore, the slow rate of breakdown of very strong protein–ligand complexes allows one to determine k_2 by incubating the radiolabeled complex with an excess of unlabeled ligand and then collecting portions of the protein–ligand complex on a semipermeable membrane as a function of time of incubation. The logarithm of the amount of radiolabeled ligand associated with the protein plotted as a function of time gives a straight line with slope equal to k_1 (69).

DETERMINING STOICHIOMETRY

Several techniques for determining stoichiometry of strong and very strong complexes were discussed earlier in Chapter 4 under the section entitled—active site concentration. These techniques require a measurement of the amount of ligand bound or the fractional saturation of a protein as a function of increasing ligand concentration. In addition to the techniques discussed in Chapter 4, the continuous variation method, also called the Job plot, can be used to determine stoichiometry. Instead of adding ligand to a fixed concentration of protein, or protein to a fixed concentration of ligand, the continuous variation method requires that both the ligand and protein concentrations be varied but that the sum total concentration of the two components be held constant. Measuring the change in some protein or ligand property that is proportional to complex formation, for example, enzyme activity or differences in absorbance or fluorescence, provides data that can be used to determine stoichiometry. When the measured parameter is plotted as a function of the mole fraction of one of the two components, a curve is obtained with a maximum that depends on the stoichiometry of the complex. If the stoichiometry is 1 mol of ligand per mole of protein, the maximum occurs at a mole fraction of 0.5. Stoichiometries greater than one cause the position of the maximum to shift to higher ligand mole fractions. However, the differences between the values of the maxima become smaller and the sharpness of the peak decreases as n increases and as $([P_T] + [L_T])/K_D$ becomes smaller. In practice then, Job's plot will give ambiguous values for the stoichiometry unless $([P_T] + [L_T])/K_D$ is large; values as large as 50 may be necessary (70). Huang (71) suggests, however, that if the position of the peak is ambiguous, the data should be replotted in terms of the mole fraction of the suspected concentration of binding sites. A symmetrical Job's plot with a maximum of 0.5 mole fraction would be consistent with one ligand bound per binding site.

THE CONFORMATIVE RESPONSE

Prior to 1950, many enzymologists considered proteins as rigid inflexible molecules. These ideas changed rapidly, however, after Koshland (72) used the induced fit hypothesis to explain enzyme substrate specificities. Basically, he provided convincing evidence that proteins are flexible molecules and that the binding of its substrate induces a new conformation. As indicated earlier then, it will be assumed that all protein molecules undergo changes in conformation after or during the binding of a ligand or when their environment is altered. Furthermore, this change in conformation is assumed to be important for the overall protein function. On the other hand, unless the change in conformation is large enough to be

Table 5.1 Methods for determining a change in protein conformation

Method
X-Ray crystallographic methods
Spectroscopic methods
Optical rotatory dispersion and circular dichroism
Ultraviolet absorption
Fluorescence methods
Reporter groups and fluorescence probes
Paramagnetic probes
Methods based on differential rates of inactivation
Thermal inactivation
Proteolytic inactivation
Chemical inactivation
Methods based on differential rates of modification
Chemical modification
Hydrogen exchange
Methods based on enzyme function
Breaks in Arrhenius plots
Transient kinetics
Allosteric kinetics
Antibody probes
Changes in sedimentation constant

detected by current technology and is related to a known function, it is of little use. For example, a single lysine residue on the surface of a protein molecule may occupy a number of positions relative to other amino acids, but the definition of each position adds little to our understanding of the protein function unless it can be correlated with some functional property.

Methodologies for detecting conformational changes are simpler than one might first imagine (73). Such procedures do not require measurement of exact structural data; it is the relative change in a property of a protein that is important. Several methods for detecting protein conformational changes are tabulated in Table 5.1. The first method listed on this table is X-ray crystallography. This is the only one that provides a full three-dimensional picture of a protein and, therefore, the only method by which one can measure the absolute change in conformation. It was first used to describe the changes in conformation of hemoglobin and lysozyme following the binding of a ligand. For a variety of reasons, however, X-ray diffraction is not routinely used in the laboratory: first, the ligand may not penetrate the crystal and gain access to all enzyme molecules; second, the protein molecule may be prevented by the forces of the crystal lattice from undergoing a change in its three-dimensional conformation; and third, special equipment and data analysis are required to complete such studies (74).

Because one or more amino acid residues change position relative to another when the conformation of a protein changes, many of these residues including parts of the peptide backbone are exposed to a different environment. For example, a tryptophanyl residue buried in the interior of the protein molecule may become exposed to the aqueous bulk solvent. A charged guanidino group may move closer to a tryptophanyl group or a carboxyl group may form a hydrogen bond with a tyrosyl phenolic hydroxyl. These changes in the environment of aromatic amino acid residues can then be detected by measuring changes in ultraviolet absorption (75), fluorescence emission (76), circular dichroism (73), and magnetic resonance (77) as discussed earlier in Chapter 4. Because the absorption spectrum of the peptide backbone is different depending on whether it is in the random coil, α helix, or β conformation, changes in the protein conformation that perturb this portion of its structure can also be detected by spectroscopic measurements (see Figure 4.4).

Spectroscopic probes (also called reporter groups), either covalently or noncovalently attached to a protein, increase the usefulness of spectroscopic measurements for monitoring changes in protein conformation (78–81). It is important to realize, though, that the attachment of a ligand by a covalent bond or the interaction of a noncovalent probe will, in itself, perturb the protein conformation. Examples of other spectroscopic measurements for detecting conformational changes in a protein are given in Reference (73).

Other methods for detecting conformational changes are based on the differential rate, or extent, of inactivation or modification of an enzyme (73). The rate of loss of enzyme activity caused by an increased temperature, and proteolytic, or chemical reactions may be perturbed by a ligand because the altered protein conformation reacts differently. In other cases, the extent of a reaction may be different when a conformation is changed. For example, the number of buried and exposed sulfhydryl groups of pyruvate kinase is different after the allosteric activator fructose 1,6-diphosphate is added (82).

Determining both the number and the rate of exchange of the peptide hydrogens with the hydrogen ions in the bulk solvent not only provides a measure of protein conformational changes but also indicates the dynamic aspects of these changes (83). The rates of exchange are monitored

Table 5.2 Responsive classes of hydrogens in hemoglobin[a]

Class	H dimer	Half-time[b] Oxy	Deoxy	$\delta\Delta G^{0c}$
Jump	3	20 sec	26 hr	4.6
	~3	20 sec	~2 hr	
Fast[d]	14	50 sec	12 min	1.5
Intermediate I	4	1.2 min	53 min	2.0
Intermediate II	12	12 min	8.8 hr	2.0
Slow I	18	106 min	50 hr	1.8
Slow II[d]	12	52 hr	2400 hr	2.1
Slow III[d]	~5	$> 10^3$ hr	$> 10^4$ hr	1

[a] Reprinted by permission from Reference (86).
[b] Exchange half-times refer to pH 7.4 and 0°C.
[c] The $\delta\Delta G^0$ value is the change in free energy of stabilization felt by each breathing unit when deoxyhemoglobin is liganded, calculated from the relationship $\delta\Delta G^0 = -RT\,\delta\ln(\text{rate})$.
[d] Rates found for the fast class at pH 6 and 0°C and for the slow II and slow III classes at pH 7.4 and 37°C are recalculated to this basis using $E_a = 20$ kcal and $d(\log k)/d(\text{pH}) = 0.7$.

principally by two methods. In the first method, the replacement of the peptide hydrogen by deuterium is followed by IR spectroscopy. As the proton is replaced by a deuteron, the absorption of the NH (amide II) band at 1545 cm^{-1} decreases (84). The second method uses the tritium isotope. First the peptide hydrogens of the protein are exchanged with tritium. To initiate the exchange reaction, protein in the tritiated aqueous solvent is added to a gel permeation column and eluted with a protiated solvent. To measure the kinetics of tritium loss, samples are taken at various times and the newly freed tritium is separated from the tritiated protein by another gel permeation column and quantified (85).

Table 5.2 shows the data for a hydrogen exchange study of oxyhemoglobin and deoxyhemoglobin (86, 87). The exchangeable hydrogens of this protein are divided into seven classes based on their rate of exchange, which for the oxy form varies from a half-time of 20 sec for the jump class to 52 hr for the slow II class. In the deoxy form, the exchange rate of each class of hydrogens becomes slower, although the magnitude of the decrease is different for each class. For the jump class, the half-time for exchange in the deoxy form is now 26 hr, over a 4000-fold decrease, while for the slow II class, it decreases by near 50-fold.

Hydrogen exchange experiments reveal something about protein conformations that other methods do not, namely, protein molecules in solution exist in a dynamic multistate conformation. Because the rate of exchange of the hydrogens in each class in oxyhemoglobin move in unison to a common slower rate when the protein is deoxygenated suggests that segments of the protein molecule undergo dynamic reversible unfolding reactions. Only during the transient unfolded state can the hydrogens in the interior of the folded molecule exchange with those in the bulk aqueous solvent (83).

Changes in enzyme function that are often correlated with conformational changes are breaks in Arrhenius plots (88), transient kinetics (89), and allosteric kinetics (73). A break in an Arrhenius plot is consistent with two different enzyme conformations, each having different energies of activation. Transient kinetics, that is, the slow response of an enzyme to a marked increase in the concentration of one of its substrates, and allosteric or sigmoid kinetic data are usually assumed to reflect changes in enzyme conformation. The altered conformer exhibits an increased or decreased K_m for one of its ligands. The slow release of a competitive inhibitor, however, can also cause transient kinetics (89).

Immunological procedures for detecting conformational changes in a protein are unique in that changes in specific parts of the peptide backbone can be monitored. Antibodies directed against a specific peptide in the protein can be used to assess the conformational equilibrium between its structured and unstructured forms (90, 91). Finally, different conformers of a protein often sediment at different rates. Using a difference sedimentation technique (92) makes it possible to measure changes of 0.2% in the sedimentation coefficients with an accuracy better than $\pm 5\%$.

ENZYME KINETICS

Enzyme kinetics is a highly developed mathematical branch of enzymology. Kinetic methods are used to determine the amount of enzyme in a tissue (see Chapter 2) and to assess some of the functional, regulatory, and mechanistic properties of an enzyme. While it may not be necessary for every enzymologist to become an enzyme kineticist, it is important that each understands and learns to use these methods. The purpose of this section is to describe the practical aspects for determining the catalytic parameters of an enzyme catalyzed reaction. A more theoretical discussion of enzyme kinetic mechanisms is available in several excellent monographs (1, 3, 4).

The Michaelis–Menten Equation

Enzymes catalyze reactions after forming complexes with their substrate as described by Equation (5.19)

$$E + S \underset{k_2}{\overset{k_1}{\rightleftharpoons}} ES \underset{k_4}{\overset{k_3}{\rightleftharpoons}} E + P \tag{5.19}$$

When the substrate concentration is considerably larger than the enzyme concentration, and when the concentration of the product is negligible so that k_4[E][P] is effectively zero, the initial velocity v of an enzyme reaction is described by the Michaelis–Menten equation (5.20)

$$v = V[S]/(K_m + [S]) \tag{5.20}$$

V is the theoretical maximum velocity and K_m is the substrate concentration at $V/2$. Note that this equation is identical in form to Equation (5.2), which describes the binding of a ligand to a protein under conditions when ligand concentrations are considerably larger than protein concentrations. Also note that $v = (V/K_m)[S]$ when the substrate concentration is much smaller than the K_m; under this condition, the reaction is first order with $k_1 = (V/K_m)$. When the substrate concentration is much larger than K_m, the equation simplifies to $v = V$.

Equation (5.20) applies to enzyme reactions involving one substrate and one product and when the plot of v versus [S] is described by a rectangular hyperbola. Most enzyme catalyzed reactions, however, involve more than one substrate and many require a cofactor as well. Yet many multisubstrate reactions can be treated as a single substrate enzyme catalyzed reaction so that the K_m values for each of the substrates can be estimated. For example, the initial velocity for a two substrate reaction in which the two substrates bind in a compulsory order is given by Equation (5.21) (3):

$$\frac{1}{v} = \frac{1}{V} + \frac{K_{mA}}{V[A]} + \frac{K_{mB}}{V[B]} + \frac{K_{mA}K_{iB}}{V[A][B]} \tag{5.21}$$

where K_{mA} and K_{mB} are the respective Michaelis constants for substrates A and B; K_{iB} is the limiting value of the Michaelis constant for B when the concentration of A approaches zero. If the concentration of B in the reaction described by Equation (5.21) is made very large, assume 100 K_{mB}, then all terms in Equation (5.21) containing [B] become negligible. It then reduces to

$$\frac{1}{v} = \frac{1}{V} + \frac{K_{mA}}{V[A]} \tag{5.22}$$

which is the reciprocal form of the equation for a single substrate enzyme reaction previously given by Equation (5.20). Determining the values for K_{iA} and K_{iB} requires a more extensive kinetic study as discussed elsewhere (1, 3, 4).

This section on enzyme kinetics will emphasize several facets of the problem of determining K_m and V: (a) How are initial velocities determined? (b) What range of substrate concentrations should be used and how should they be spaced? and (c) How should the data be analyzed?

Determining Initial Velocities

Practically every enzymologist uses manual methods to determine the initial velocity of an enzyme reaction. For an end point assay, the initial velocity is the amount of product formed in a reaction divided by the elapsed time. A valid measurement of this velocity requires a constant rate of reaction over this time period and carefully controlled reaction conditions as discussed earlier in Chapter 2 or in Reference (93). It must be realized, however, that initial velocity measurements discussed in Chapter 2 have different objectives than those discussed here. Those discussed in Chapter 2 were designed to determine the amount of enzyme in a sample. Optimal pH, substrate, and activator concentrations were required to produce a routine, reproducible, and sensitive assay. Determining K_m and V from initial velocities, on the other hand, require that measurements often be made when conditions are not optimal. Substrate and activator concentrations are often not saturating and other conditions, such as pH, may not be optimal.

Loss of the activity of the diluted stock enzyme solution as a function of time must be monitored when collecting a set of v and [S] data pairs. Recording the time each velocity measurement is made makes it possible

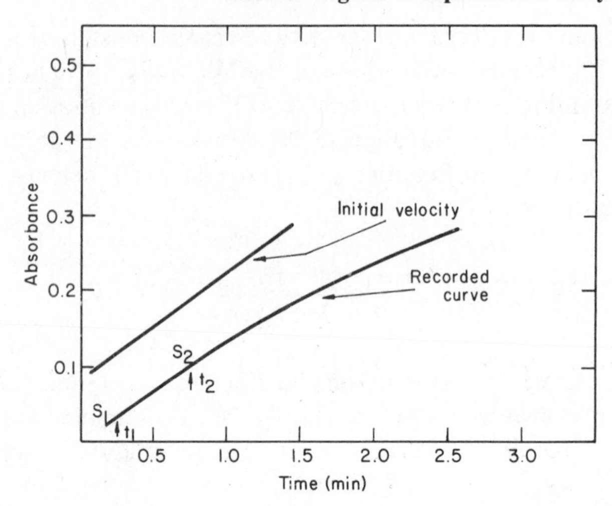

Fig. 5.8. Portion of a hypothetical enzyme reaction progress curve. The line labeled "initial velocity" is drawn using a transparent ruler with a fine line etched on it as a tangent to the initial part of the recorded curve. $[S]_1$ and $[S]_2$ are substrate concentrations at times t_1 and t_2, respectively.

to correct for loss of the stock enzyme activity. For careful work, the activity of the stock enzyme should be monitored several times during the experiment using two substrate concentrations, one at 10 times K_m and one at the K_m value. If the rates of loss of activity at the two substrate concentrations are identical, then the values of v can be corrected accordingly. If the loss of activity at K_m concentrations of substrate is more rapid than at the higher substrate concentrations, then the K_m for the enzyme is not stable with time and the data set should be discarded.

A correct K_m value also requires that the substrate concentration be known accurately. It is not a good practice, however, to rely on the purity of the reagent indicated on its container. Determining its concentration enzymically ensures that the K_m will be based on the enzymically active form of the substrate.

If the reaction is monitored with a chart recorder, as preferred by most investigators, the slope of the tangent to the recorded reaction curve at time equal to zero represents the initial velocity. Using a transparent ruler with a fine line etched on it parallel to the edge makes it possible to draw a tangent without defacing the recorded curve (see Figure 5.8). Alternatively, a small front surfaced mirror placed upright on the tracing at right angles to the curve is a useful aid for drawing the tangent (94). The mirror is moved until the reflection of the recorded curve on the mirror is continuous with the curve itself. A line drawn along the mirror surface in contact with the paper is exactly perpendicular to the curve; a tangent to the recorded curve will cut this line at a 90° angle.

Drawing the tangent to the recorded curve is easy, particularly if the initial trace is almost linear. If the recorded trace is appreciably curved or noisy, it becomes more difficult. Because of this and because manual methods tend to give low estimates of v, even when the recorded trace is almost linear, several investigators have described other methods for determining v (95).

Lee and Wilson (96) and Waley (97) determine the initial velocity from the slope of a chord joining $[S]_1$ and $[S]_2$ on a reaction progress curve as shown in Figure 5.8. The slope of the chord joining the points is the initial velocity at a substrate concentration given by Equation (5.23) (97).

$$[S] = \frac{([P]_2 - [P]_1)}{\ln(([S]_0 - [P]_1)/([S]_0 - [P]_2))} \qquad (5.23)$$

If loss of substrate is monitored rather than formation of product, the slope of the chord gives the initial velocity at a substrate concentration [S] given by

$$[S] = ([S]_1 - [S]_2)/\ln([S]_1/[S]_2) \qquad (5.24)$$

Instead of using Equations (5.23) and (5.24) to calculate the [S] value corresponding to the initial velocity given by the slope of the chord, Lee and Wilson (96) approximate v and [S] by the following equations:

$$v = ([S]_2 - [S]_1)/\Delta t \qquad [S] = ([S]_1 + [S]_2)/2$$

The error in Lee and Wilson's approximation of the initial velocity is only 4% when $[S]_2$ is one half the initial substrate concentration. Because Lee and Wilson's method is not much simpler than the exact method, particularly if a programmable hand held calculator is available, it is preferable to use the exact method for calculating [S].

Alternative methods for determining initial velocities use the direct linear plot (98), computer assisted fits of data taken from the initial velocity curve to the integrated Michaelis–Menten equation (95), or fits to polynomials (95). Of these three methods, the direct linear plot appears to be the method of choice (98). Determining initial velocities by alternative methods, however, offers little advantage over manual methods and will not be discussed.

Designing a Kinetic and Binding Experiment

The major decisions that must be made when designing kinetic and binding experiments are: What range of ligand and protein concentrations should be employed, and how many different concentrations of each should be used?

Using the formula for the propagation of errors applied to Equation (5.1), and assuming that the errors in $[P_T]$ and $[L_T]$ are negligible compared to the error in determining δ or $\overline{v}$, it is possible to show that the most accurate values of K_D are obtained from data collected when fractional saturation ranges from 0.2 to 0.8; the most accurate values for δ_{max} or n are obtained above 80% saturation (99). These calculations are depicted graphically in Figure 5.9.

The range of substrate concentrations used for a kinetic determination of K_m and V is based on the same criteria as that used to design a binding experiment. The dependent variable, the initial velocity in the enzyme kinetics experiment, provides the most accurate values for K_m when the substrate concentrations range from 20 to 80% saturation (Figure 5.9).

The spacing of the ligand concentrations is also important because it has a strong influence on the efficiency of the statistical methods used to analyze them (100): the higher the efficiency, the smaller the variance (101). For a weighted least squares fitting of the data, the ligand concentrations should be spaced geometrically, that is, successive concentrations should be in a constant ratio, for example, 1:2:4:8 (102). Cleland (103) recommends using substrate concentrations equal to 0.2 K_m, 0.5 K_m, K_m, $2K_m$, and $5K_m$ while Henderson (104) recommends values equal to $K_m/4$, $K_m/3$, $K_m/2$, K_m, $2K_m$, $4K_m$, and $8K_m$. If the nonparametric method is used to analyze the data, the substrate concentration should be equally spaced not geometrically spaced (100). If there is any question about the correctness of the equation used to fit the data, it is best to use a minimum of 10 data pairs. As discussed later in this chapter, determining whether an equation describes a data set requires a statistical analysis of the randomness of the residuals around the fitted line. The decision is made with more confidence when 10 or more data pairs are used. However, 10 substrate concentrations cannot be spaced geometrically as in-

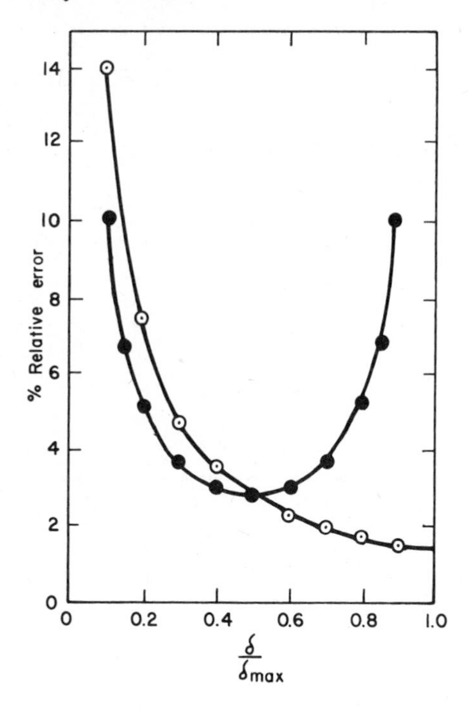

Fig. 5.9. Percent relative errors $100\,\Delta K/K$ (●) and $100\,\Delta\epsilon/\epsilon$ (○) as a function of the fractional saturation, δ/δ_{max}. The curves were calculated with Equation (6, ●) and Equation (7, ○) of Reference (99) assuming a constant $(\Delta\delta/\delta_{max}) = 0.01$.

dicated above. Therefore, I recommend that subsequent concentrations in a data set have a ratio of 1.6 instead of 2.

Because the ideal range of substrate concentrations needed to determine the kinetic or binding parameters is expressed in terms of a K_m or K_D value, a trial experiment is needed to estimate it. Determining δ, v, or $\overline{v}$ from the median value of five different measurements at each of two concentrations—one at high ligand concentration and one at low ligand concentration—and plotting them in the direct linear plot gives a good estimate of the parameters (105). Instead of plotting the two median measurements in the direct linear plot, the constants can be calculated using Equations (5.25) and (5.26)

$$V = ([S]_2 - [S]_1)/(([S]_2/v_2) - ([S]_1/v_1)) \tag{5.25}$$

$$K_m = (v_2 - v_1)/((v_1/[S]_1) - (v_2/[S]_2)) \tag{5.26}$$

As pointed out earlier in this chapter, determining the K_D or K_m of weak complexes requires that $n[P_T]/K_D$ be small, so that an insignificant

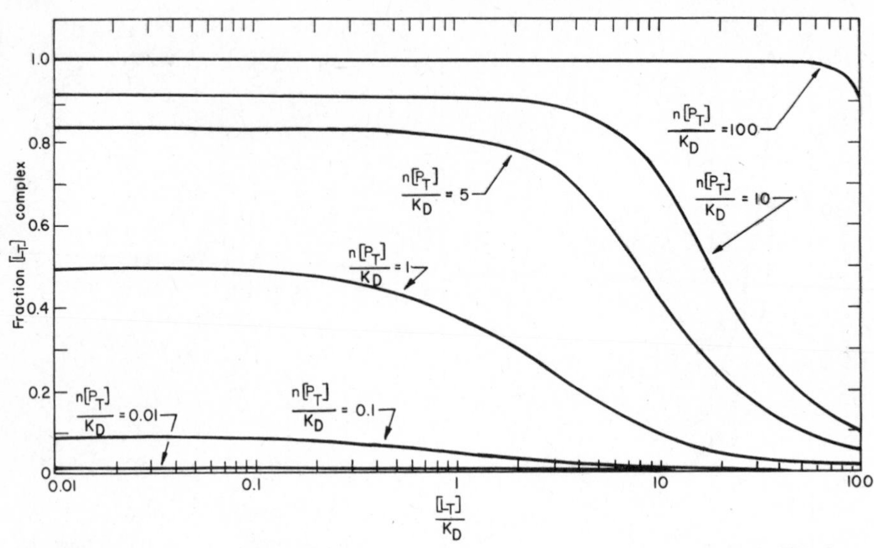

Fig. 5.10. Fraction of the total ligand in a reaction complexed by a protein, $[L_B]/[L_T]$, as a function of $[L_T]/K_D$ at various protein concentrations $n[P_T]/K_D$. $[L_B]$ was calculated with Equation (5.1).

amount of ligand is complexed at any point in the experiment. (If ligand is titrated with protein, then $[L_T]/K_D$ must be small so that an insignificant fraction of the protein binding sites are complexed.) Under these conditions, the concentration of free ligand is approximately equal to the concentration of total ligand ($[L] = [L_T]$ or $[S] = [S_T]$ and $[L_B]/[L_T]$ is negligible for all values of $[L_T]$. A graphical view of this statement is portrayed in Figure 5.10. Plotting the fraction of the total ligand that is bound in a complex as a function of $[L_T]/K_D$ for various protein concentrations shows that for $[L_B]/[L_T]$ to be below 0.02, $n[P_T]/K_D$ must be near 0.01.

The question of interest, however, is how large can $n[P_T]/K_D$ be before there is an appreciable error in the calculated constants. An extensive discussion of this problem is beyond the scope of this chapter. However, fitting simulated data with a weighted least squares regression, as discussed later in this chapter, shows less than 5% error in the kinetic or binding parameters when $n[P_T]/K_D = 0.1$. Because the error increases dramatically when $n[P_T]/K_D$ values are greater than 0.1, I recommend using protein concentrations that give smaller values.

ANALYZING KINETIC AND BINDING DATA

Analyzing kinetic and binding data is basically a two step process. The first step, sometimes called model building, gives an indication of the mechanism of the reaction: Calculating the kinetic and binding constants makes up the second step. When the main objective of an enzyme kinetic and binding experiment is to determine the underlying reaction mechanism, considerable time and effort may be expended in the model building phase of the process. For this discussion, however, the main objective is to obtain the relevant binding or kinetic constants. Mechanisms or models more complex than those described by the Michaelis–Menton equation or the binding function [Equation (5.23)] will not be discussed.

Modeling enzyme kinetic and binding data is often done by graphical methods. Most investigators plot data in the linear form because deviations from linearity are easier to visualize than deviations from a curved line. Before computers became available, this same linear plot was used to determine the relevant kinetic or binding constants. This common practice of the past, as still portrayed by many basic textbooks of biochemistry, of manually calculating the relevant parameters by anyone of the linear transformations of the Michaelis–Menten or binding equation, is no longer acceptable. Using the computer to calculate the constants is more objective and precise. The graphical presentation of the data is then used to show that the correct model is used in the calculation.

Whether or not a set of data is described by a specific model or equation is often determined by examining the graphical plot by eye. Many investigators simply draw a theoretical line through the plotted data using the calculated binding or kinetic constants. If the majority of the experimental points appear to fall on the line, it is often assumed that the correct model can be used to calculate the constants. Mannervik (102) suggests a more objective procedure for making this decision based on the randomness of the residuals q_i. Residuals are the differences between the observed values of the dependent variable and the corresponding calculated values. For initial velocities,

$$q_i = v_i - \hat{v}_i \tag{5.27}$$

where $\hat{v}_i$ is the calculated value obtained by regression analysis. The residuals of a good fit are expected to be distributed about the zero value with approximately equal numbers of positive and negative values. The randomness of the residuals can be tested by ordering residuals in a linear array. For example, assume that the residuals have the following groupings of signs as a function of increasing substrate concentrations

$$(+\,+\,+)(-)(+\,+)(-)(+\,+)(-\,-)(+)(-)\,(+)(-)(+)$$

The negative sign appears 6 times in this sequence, the positive sign 10 times. The probability is 0.958 that 11 different runs of alternating positive and negative signs with 6 negative and 10 positive signs is random. Therefore, the residuals in this calculation have a random distribution. On the other hand, if the residuals have the following grouping

$$(-)(+\,+)(-)(+\,+\,+\,+\,+\,+\,+\,+)(-\,-\,-\,-)$$

where the negative and positive signs again appear 6 and 10 times, respectively, but in 5 runs instead of 11, the probability that the grouping is random is 0.047. The fit of the data to this model is inadequate at the 95% confidence level. Testing the randomness of other arrays of residuals is conveniently done by using the set of numbers provided in Appendix III.

Several linear transformations of the equations that describe binding and kinetic data are possible and some of their graphical equivalents are shown in Figure 5.11. The double reciprocal plot, called the Benisi–Hildebrand plot when binding data are plotted, and the Lineweaver–Burk plot, when enzyme kinetic data are plotted, is presented in Figure 5.11(*a*). It is a popular plot, and particularly useful in testing the correctness of an enzyme kinetic mechanism (106). As will be discussed later, enzyme inhibition data are easy to interpret when plotted in the reciprocal format. Its major limitation is that the abscissa is unbounded so that any range of values of the ligand concentration can be represented, creating an unfair impression of the accuracy and range of the data.

The direct linear plot [Figure 5.11[*b*)] is useful for manually estimating K_D, K_m, V, or n as noted earlier, but it does not give a clear picture of the range of the data or the correctness of the mechanism. Figure 5.11(*c*), the Hanes plot, provides a more faithful reflection of the accuracy of the initial velocities than the reciprocal plot. It is preferred by some investigators (1). Figure 5.11(d) is called a Scatchard plot when binding data

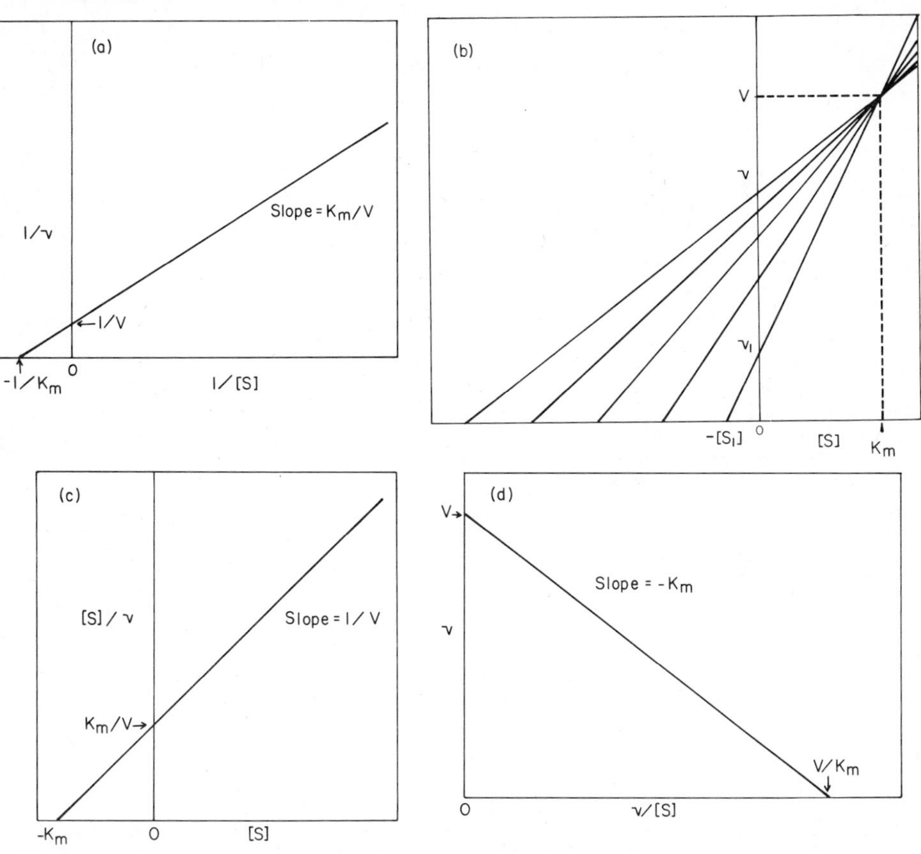

Fig. 5.11. Graphical equivalents of Equations (5.2), (5.10), or (5.20). (*a*) Benisi–Hildebrand or Lineweaver–Burk plot; (*b*) Direct linear plot; (*c*) Hanes plot; and (*d*) Eadie–Hofstee or Scatchard plot.

are plotted and a Eadie–Hofstee plot when enzyme kinetic data are plotted. It, like the double reciprocal plot, also compresses data at high ligand concentrations and creates an unfair impression of the range of the data. Klotz and Hunston (107) discuss published examples of binding data that were analyzed by the Scatchard plot to give values for the stoichiometry of a reaction. These were consistent with literature values current at the time but which later turned out to be in error, because the proper range of ligand concentrations were not used. For example, Figure 5.12 shows hypothetical data presented in terms of both a Scatchard plot and the nonlinear Klotz plot. The Klotz plot clearly indicates the weakness of the

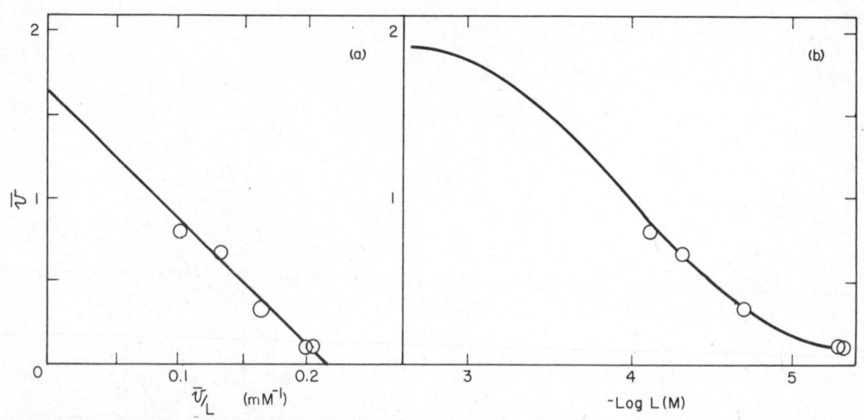

Fig. 5.12. Scatchard (*a*) and Klotz plot (*b*) of data for the binding of ligands to a protein containing two identical noninteracting binding sites. Insufficient data for determining binding constants when plotted in the Klotz format show the weakness of the Scatchard presentation.

Scatchard presentation: Binding data using ligand concentrations that were not sufficient to achieve 80% saturation were extrapolated to give incorrect values of n.

Ferguson–Miller and Koppenol (108), on the other hand, show that the Scatchard or Eadie–Hofstee plot is useful because it provides a more sensitive indication of the complexity of the binding system and the extent of deviation from linearity than the Klotz plot. They examined four different cases. Case one involves a protein with two identical, noninteracting sites having equal site binding constants. The Klotz plot for such a system, Figure 5.12(*b*), gives a sigmoidal curve with a single inflection point. The value of [L] at the inflection point corresponds to the site binding constant K_D. As the concentration of L increases, the curve asymptotically approaches the number of binding sites n. The Scatchard plot of the data [Figure 5.12(*a*)] gives a straight line with a slope corresponding to $1/K_{D1}$ and an intercept to n.

In the second case, described by

$$\bar{v} = [L]/(K_{D1} + [L]) + [L]/(K_{D2} + [L]) \tag{5.28}$$

the two binding sites are noninteracting with widely different ligand affinities: $K_{D1} \ll K_{D2}$, $K_{D1} = K_{D3}$, and $K_{D2} = K_{D4}$. In this case, the Scatchard plot is strongly biphasic [Figure 5.13(*a*)]; the slope of the first phase is $1/K_{D1}$ and for the second phase $1/2K_{D2}$. The same data plotted as a Klotz

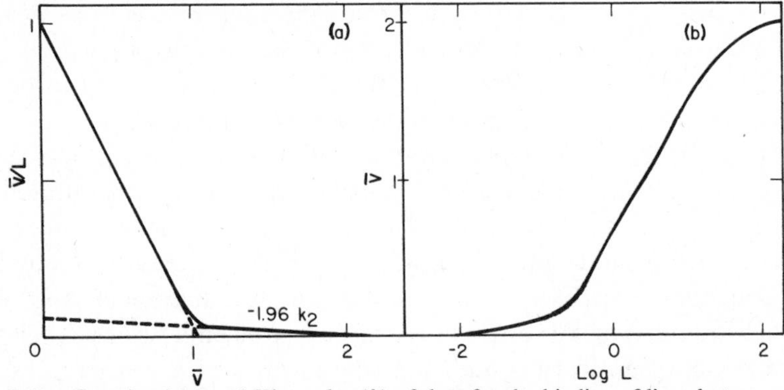

Fig. 5.13. Scatchard (*a*) and Klotz plot (*b*) of data for the binding of ligands to a protein molecule with two binding sites whose affinites for ligand differ by 50-fold. The site constants were assigned the numerical values of 1 and 50, so that $K_{D1} = K_{D4} = 1$ and $K_{D2} = K_{D3} = 50$ and hence the stoichiometric constants are $K_{St1} = 0.98$ (approximately K_{D1}) and $K_{St2} = 51$ (approximately K_{D2}) (*a*) A Scatchard plot of the data in which the *y*-intercept of the first phase is $1/K_{D1}$. The slope of the first phase is $-1/K_{D1}$, the second phase $-1/2K_{D2}$. (*b*) A Klotz plot of the data. Abstracted with permission from S. Ferguson–Miller and W. H. Koppenol, *Trends Biochem. Sci.*, **6**, 1–3 (1981). Reference (108).

plot [Figure 5.13(*b*)] show a decreased slope but not two distinct inflection points.

The third and fourth cases that were examined are more complex. In each case though, the Scatchard or Eadie–Hofstee plot is a more sensitive indicator of the deviation from linearity and the correctness of the mechanism. The Klotz plot, however, gives a better visual display of the range of ligand concentrations employed in the experiment.

Deciding on what plot to use depends on the objective of the experiment. When reaction data are known to fit the Michaelis–Menten equation or the binding function, then as indicated earlier, the constants for the reaction should be calculated by mathematical procedures. A graphical illustration of the data is then used to show both conformity to a model and the range of ligand concentrations employed in the experiment. If the reaction is more complex, and the mechanism or model is not known, then it may be necessary to estimate reaction parameters from the intercepts and slopes of the plotted data.

Assuming that a set of binding or kinetic data is adequately modeled by Equations (5.2) and (5.20), respectively, the next step in the process is to calculate the relevant constants. Before the data are fitted to the proper equation, however, one needs to know the structure of the errors in the data so that the proper statistical procedures are employed. As indicated earlier, however, the spacing of the data points in an experimental design should be based on the statistical procedures to be employed in analyzing the data. In actual practice then, the potential errors in a data set should be appreciated before the experiment is planned.

Because a skilled experimentalist has little difficulty in accurately defining the substrate concentrations in a reaction mixture, the major errors in an enzyme binding or kinetic study occur during the measurement of the dependent variables, v, $\bar{v}$, or δ. The question that remains is: What is the structure of the error in these measurements? Does it follow a normal Gaussian distribution? Does the error have a constant absolute magnitude or does it vary with the magnitude of the dependent variable in a constant proportional manner?

The problem is that most investigators do not take the time or make the effort to complete the number of measurements required to define the error in a binding or kinetic study. Although the few attempts that have been made have given somewhat different results, the consensus among most investigators is that the error follows a normal Gaussian distribution and that the absolute value of the error in the dependent variable increases with its magnitude (109). In other words, the relative error is constant. Such kinetic or binding data should be fitted with a weighted least squares procedure.

Some investigators, on the other hand, feel there is insufficient evidence to support a constant relative error in the dependent variable. Primarily championed by Cornish-Bowden (1), they argue that enzyme kinetic data should be analyzed by a method that is little affected by the structure of the error. The preferred method in this case is the direct linear plot (100) because it includes fewer assumptions about experimental error than the least squares regression analysis. Those recommending the use of the direct linear plot argue that initial velocities which deviate considerably from predicted values (called outliers) have very little influence on this analysis but have a profound influence on the least squares regression analysis.

The direct linear plot, however, has not achieved wide use. As emphasized by Mannervik (102), it is not safe to effectively eliminate outliers, as the direct linear plot does, because they may reflect a different model or the true error of the data set. Other difficulties detracting from the usefulness of the direct linear plot are (a) the calculated standard deviations or confidence limits of the kinetic or binding constants are only approximations, and (b) the plot of the data offers little insight into the correctness of the kinetic or binding equation.

At least two or preferably three determinations of v should be made at each substrate concentration. Replicate measurements give an indication of the precision of the data and also help in deciding whether a point might be an outlier. Do not average the replicate measurements but rather submit all points for analysis.

The preferred procedure for calculating binding and kinetic constants is to fit the data directly to the proper equation by regression analysis (102, 110). The weighted linear regression procedure, although tedious, is simple if done manually (1, 111). Using computers and the appropriate software makes this procedure more practical. Therefore, anyone who anticipates analyzing initial velocity or binding data should take the time to set up the proper statistical procedures. Many computer programs are available (110, 112–119). The application of computer programs to enzyme kinetic data are reviewed in Reference (120). Wilkinson's manual method is given in Appendix IV.

The weighted linear regression method given in Appendix IV assumes that the constant error has a normal Gaussian distribution absolute. The dependent variable is weighted, which in effect means that the fitting is equivalent to an unweighted regression to an enlarged set of points, each point of the original set being repeated the appropriate w (weighting) times. In this way, the more accurate higher values of the dependent variable are given more weight. The weighting factors are derived from the weighting function

$$w_i \propto v_i^{-\alpha} \tag{5.29}$$

where $\alpha = 0$ for a constant absolute error and 2 for a constant relative error. Residual analysis of actual enzyme kinetic data gives a value somewhat lower than 2 (121, 122). For simplicity though, most investigators assume $\alpha = 0$. Instead of using a fixed value for α, Mannervik (102) recently suggested that this value should be determined for each subgroup of 5 or 6 residuals in a data set. Obtaining the weighting factors from local variances in a data set will account for differences in the error structure of each experiment.

If an outlier appears to bias the least squares regression analysis, it may be eliminated if the residual of that point exceeds 2σ. Mannervik (102) defines σ by

$$\sigma = \sqrt{SS/(n - p)}$$

where SS is the sum of the squares of the residuals defined by Equation (5.27), n the number of data pairs, and p the number of parameters in the model. For the Michaelis–Menten model, $p = 2$. Mannervik (102) emphasizes that only one point at a time and only very few data, if any, should be eliminated from a data set. If several points deviate, then the equation used to fit the data may be wrong or other problems may have been encountered such as loss of enzyme activity with time or substrate inhibition.

The initial velocity method for analyzing the kinetics of enzyme reactions was developed because early workers in enzyme kinetics encountered many difficulties when integrated rate equations were used. However, monitoring enzyme reactions using computer assisted procedures for data collection and analysis (123) makes it worthwhile today to consider analyzing enzyme reaction progress curves for K_m and V. In principle, it should be possible to obtain K_m and V from a single reaction progress curve. In practice, however, a number of progress curves are needed for a valid interpretation of the data. In addition, this methodology has its pitfalls; several precautions must be observed.

One of the several forms (124) of the integrated Michaelis–Menten equation is given by

$$([S]_0 - [S])/t = V - K_m(1/t)\ln([S]_0/[S]) \qquad (5.30)$$

where $[S]_0$ is the initial substrate concentration and [S] is the substrate concentration remaining at time t. A plot of $([S]_0 - [S])/t$ versus $(1/t)\ln([S]_0/[S])$ should give a straight line with slope of $-K_m$ and intercepts on the vertical and horizontal axis of V and V/K_m, respectively. Alternatively, v at several substrate concentrations along the progress curve can be measured manually and used to calculate V and K_m by the standard initial velocity methods (125, 126). The differential equations that describe a reaction progress curve can also be solved numerically and the calculated curve can be compared with the experimental points by eye (127).

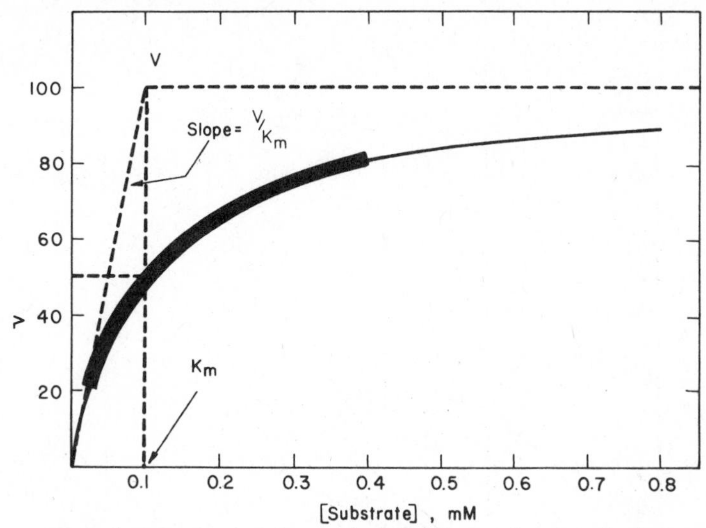

Fig. 5.14. Michaelis–Menten plot of enzymic velocities as a function of substrate concentration. Abstracted with permission from Reference (128).

This procedure provides estimates of K_m and V but gives no information on the reliability of the estimates.

Two of the major difficulties encountered when determining K_m and V from a complete reaction progress curve are product inhibition and loss of enzyme activity during the reaction. To check for product inhibition, determine K_m and V from progress curves starting at different substrate concentrations. The same values for K_m and V should be obtained at all substrate concentrations. If the enzyme loses activity during the reaction, the K_m and V values will not be the same when two different concentrations of enzyme are used to generate complete reaction progress curves (124).

Determining V and V/K_m

Rewriting the Michaelis–Menten equation with V and V/K_m as the primary constants gives

$$v = V[\mathrm{S}](V/K_m)/(V + [\mathrm{S}](V/K_m)) \tag{5.31}$$

As shown in Figure 5.14, V and V/K_m are, respectively, the asymptote at high [S] and the slope of the tangent at low [S] of the curve of v versus

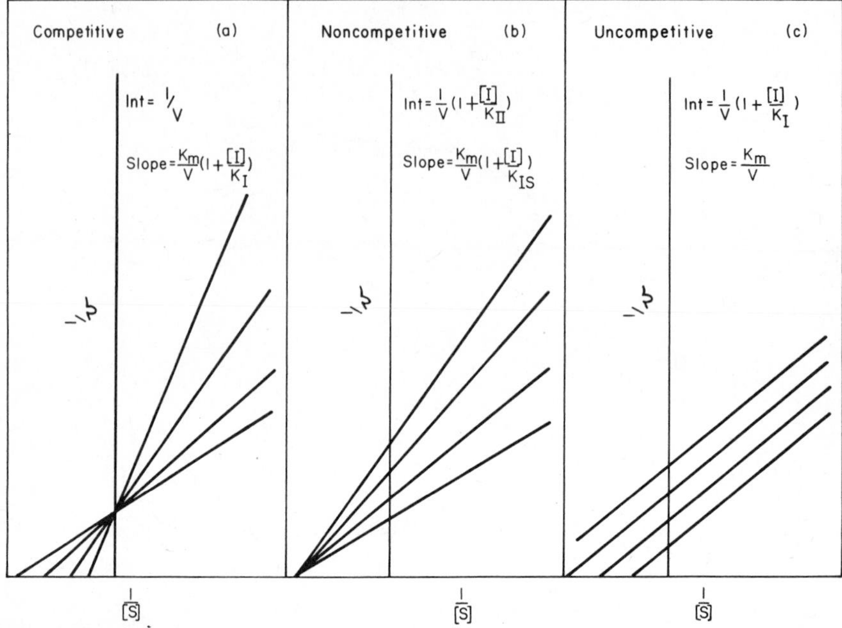

Fig. 5.15. Graphical equivalents of enzyme inhibition kinetics as defined by Equations (5.32), (5.33), and (5.34).

[S] (128). Note that the values of V/K_m and V are not dependent on each other. Only the K_m value is influenced by both the slope of the tangent and the position of the asymptote. Therefore Northrup (128) suggests fitting enzyme kinetic data to Equation (5.31) rather than Equation (5.20). Similar arguments were made earlier by Cleland (129) and thus printouts from his computer programs for analyzing enzyme kinetic data provide values for these constants (110).

Enzyme Inhibition

Three types of enzyme kinetic inhibition patterns are commonly observed: competitive, noncompetitive, and uncompetitive. These are described by Equations (5.32), (5.33), and (5.34), with their graphical equivalents given in Figure 5.15.

$$1/v = 1/V + (K_m/V[S])(1 + ([I]/K_I) \tag{5.32}$$

$$1/v = (1/V)(1 + ([I]/K_{II}) + (K_m/V[S])(1 + [I]/K_{IS}) \tag{5.33}$$

and

$$1/v = (1/V)(1 + [\mathrm{I}]/K_{\mathrm{I}}) + K_{\mathrm{m}}/V[\mathrm{S}] \tag{5.34}$$

A visual examination of these equations written in their reciprocal form shows that for competitive inhibition, the slope term is modified by $(1 + [\mathrm{I}]/K_{\mathrm{I}})$; the intercept, $1/V$, is not modified [Figure 5.15(*a*)]. Therefore, the substrate concentration at $V/2$, that is, the apparent K_{m}, increases as the inhibitor concentration is increased; $K_{\mathrm{m}}^{\mathrm{app}} = K_{\mathrm{m}}(1 + [\mathrm{I}]/K_{\mathrm{I}})$. It is important to remember that only the apparent K_{m} changes. The actual K_{m} value remains unchanged.

The equation for noncompetitive inhibition [Equation 5.33)] shows that both the slope and intercept terms of the reciprocal plot are affected by the inhibitor [Figure 5.15 (*b*)];

$$K_{\mathrm{m}}^{\mathrm{app}} = K_{\mathrm{m}}(1 + [\mathrm{I}]/K_{\mathrm{IS}}) \quad \text{and} \quad 1/V^{\mathrm{app}} = 1/V(1 + [\mathrm{I}]/K_{\mathrm{II}}).$$

If $K_{\mathrm{II}} = K_{\mathrm{IS}}$, the inhibition is simple noncompetitive. If $K_{\mathrm{II}} \neq K_{\mathrm{IS}}$, then the inhibition is mixed or partially competitive. Figure 5.15(*b*) shows simple noncompetive inhibition, the extrapolated lines all meet on the ordinate left of the origin. When $K_{\mathrm{II}} \neq K_{\mathrm{IS}}$, the extrapolated lines intersect at a point above or below the ordinate in the third or fourth quadrant.

Uncompetitive inhibition is characterized by parallel reciprocal plots [Figure 5.15(*c*)]. The inhibitor affects V and K_{m} equally, V/K_{m} is not affected. This type of inhibition is commonly observed when inhibitors are used to investigate the kinetic mechanism of multiple substrate enzyme catalyzed reactions (1, 3, 4). It will not be discussed further.

While the different patterns of inhibition are relatively easy to visualize in terms of the three equations given above, the mechanism of the inhibition process is not always easy to decipher (1, 3, 4, 130, 131). In the simplest case, a competitive inhibitor and substrate compete with each other for binding at the active site of the enzyme. A number of cases have been described, however, where competitive inhibition is observed but inhibitor and substrate bind at distinct sites (130). Furthermore, not all competitive inhibitors are structurally similar to the substrate. For example, salicylate is a competitive inhibitor of many dehydrogenases; it competes with the nucleotide for the adenine binding site.

Noncompetitive inhibition is more complex. Equation (5.33) is derived assuming inhibitor binds to both the free enzyme and the enzyme–substrate complex with binding constants K_{IS} and K_{II}, respectively; as indicated earlier, K_{IS} and K_{II} may or may not be identical. When $K_{II} = K_{IS}$, only V^{app} is affected. If $K_{II} \neq K_{IS}$ both V^{app} and K_m^{app} are changed. In addition, some compounds behave as simple noncompetitive inhibitors, but do not form reversible complexes with the enzyme. They irreversibly inactivate a portion of the enzyme. By doing so V is changed, but the remaining enzyme still exhibits the same K_m. In the strictest sense, they should not be called noncompetitive inhibitors.

Because the inhibition complexes discussed above are weak complexes, Equations (5.32), (5.33), and (5.34) are derived assuming that an insignificant amount of the added inhibitor is bound, that is, $[I] \gg [E]$. When very strong binding inhibitors, also called tight binding inhibitors, are encountered, however, the equations given above do not apply. First, very strong binding inhibitors usually have K_I values in the concentration range of the enzyme in the kinetic assay. A significant portion of these inhibitors may be complexed with the protein so that the concentration of the free inhibitor is no longer equal to the total concentration of inhibitor added to the reaction mix. Furthermore, the initial rate measurements do not have the same meaning if the rate of a reaction between an inhibitor and an enzyme is slow. If a reaction is started by adding enzyme to a reaction mixture containing both substrate and a low concentration of a very strong binding inhibitor, the substrate may be converted to product before a significant amount of the enzyme–inhibitor complex forms, in fact, inhibition may not be observed. On the other hand, if a very strong binding competitive inhibitor is preincubated with the enzyme for sufficient time to achieve equilibrium and then the reaction started by adding substrate, the initial velocity measurement would reflect the amount of free enzyme remaining because the inhibitor would be displaced slowly by the substrate; a plot of $1/v$ versus $1/[S]$ at different inhibitor concentrations yields a noncompetitive inhibition pattern similar to Figure 5.15(*b*). Therefore, the study of very strong binding inhibitors is more complex than weak or strong binding inhibitors.

One useful technique for examining the interaction of very strong binding inhibitors is to preincubate various amounts of the enzyme with a fixed concentration of inhibitor for sufficient time to establish equilibrium and then to assay an aliquot of the mixture for enzyme activity. Plotting v versus enzyme concentration $[E_T]$ gives a plot depicted in Figure 5.16, called the Ackerman–Potter plot (132). If the inhibitor competes with the substrate for the catalytic site, the enzyme in the EI complex will not

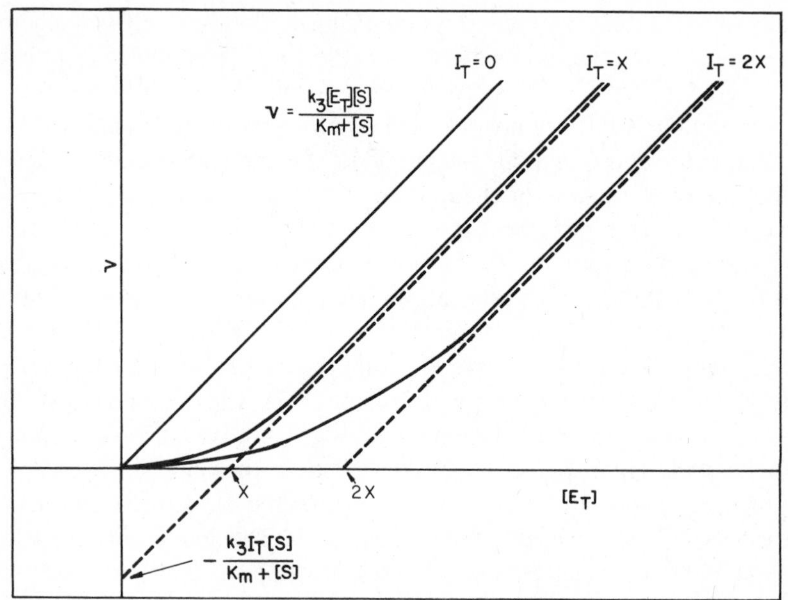

Fig. 5.16. Ackerman–Potter plot illustrating the effect of a competitive very strong binding inhibitor on the initial velocities of an enzyme reaction.

contribute to the initial enzymic reaction velocity because it dissociates slowly. As shown in Figure 5.16, at low concentration all of the enzyme exists as an EI complex and thus does not exhibit activity. Only when its concentration exceeds the inhibitor concentration is the slope of the plot directly proportional to enzyme concentration. As the enzyme concentration is increased, the plot approaches an asymptote as defined by

$$v = (k_3[\mathrm{S}])([\mathrm{E_T}] - [\mathrm{I_T}])/(K_\mathrm{m} + [\mathrm{S}]) \quad (5.35)$$

The asymptote intersects the $[\mathrm{E_T}]$ axis at $[\mathrm{I_T}] = [\mathrm{E_t}]$ and the v axis at $v = -k_3[\mathrm{I_T}][\mathrm{S}]/(K_m + [\mathrm{S}])$. Very strong binding inhibitors are of greater interest today than in the recent past because of the development of transition state analogs (133), slow tight binding inhibitors (134), and mechanism based inactivators (135).

Enzyme Activation

Many enzymes require cofactors for catalysis. Some cofactors, like the purine and pyrimidine nucleotides, are cosubtrates in that they are chemically modified and released as products during catalysis. Other cofactors, like biotin, pantetheine, and some metallo ions, are covalently attached to the enzyme; most of these cofactors also undergo covalent modification during catalysis but are not released as products of the reaction.

Enzymes requiring monovalent and/or divalent cations either bind the substrate or the cation or their complex as the first step of the catalytic process. In some enzymes, the cations have a dual role. For instance, pyruvate kinase requires 1 mol of a monovalent and 2 mol of a divalent cation per catalytic site (136). Monovalent cations activate by forming a complex with the enzyme. One mole of the divalent cation interacts with the enzyme to form the active enzyme, the other forms a complex with the nucleotide to form the active substrate (137). The data show that Mn^{2+} is a better activator of the enzyme than Mg^{2+} but that the Mg^{2+} complex of adenosine diphosphate is a better substrate than the Mn^{2+} complex.

In most cases, the activation of an enzyme by an activator can be described by the Michaelis–Menten equation. The activation constant K_A for the cofactor is determined from initial velocity studies as will be outlined later. Cofactors that act as cosubstrates, on the other hand, are treated as substrates. Of course, covalently linked cofactors are considered as part of the enzyme. Because the activation of enzymes by metal cofactors is more unique, it will be the major focus for much of the remainder of this section.

Some metalloenzymes, like Zn^{2+} metalloenzymes, bind their cofactors with very high affinity, that is, they form very strong complexes. Because the K_A for these complexes is difficult to determine, it is their stoichiometry which is most often investigated. Adding increasing amounts of the cofactor and plotting the activity as a function of the concentration of the cofactor gives a plot similar to Figure 4.1 from which it is possible to calculate stoichiometry.

Determining the activation constants for cofactors that form strong metallocomplexes is, in principle, identical to determining the K_m for a substrate. Initial velocities are determined at a fixed substrate concentration as a function of increasing concentration of the cation. If the enzyme expresses no activity in the absence of the cation, the v and [L] data pairs are analyzed as previously discussed to obtain K_A. If the enzyme expresses partial activity in the absence of cations, this partial activity must be subtracted from each velocity measurement before the data can be analyzed. Whether the K_A reflects the binding of the cation to the substrate or to the enzyme or the binding of the substrate–cation complex to the enzyme, however, requires additional study. The question of interest is: Is the divalent cation–substrate complex the true substrate or is the enzyme–cation complex the active form of the enzyme?

Consider the following reaction,

$$\begin{array}{ccc} S + M & \underset{}{\overset{K_D}{\rightleftharpoons}} & SM \\ & & + \\ & & E \\ & & \Big\Updownarrow K_{SM} \\ & & ESM \rightleftharpoons E + P + M \end{array}$$

where M is the cation activator. Substituting [SM] from Equation (5.36),

$$K_D = [S][M]/[SM] \tag{5.36}$$

into the Michaelis–Menten equation and rearranging gives

$$v = V[S][M]/(K_{SM}K_D + [S][M]) \tag{5.37}$$

Equation (5.37) can be rearranged into two equivalent equations

$$v = V[S]/((K_DK_{SM}/[M]) + [S]) \equiv V[M]/((K_DK_{SM}/[S]) + [M])$$

These two equivalent equations show that the initial velocities behave the same when plotted as a function of either free S or M. Because the equilibrium concentration of free S at some fixed concentration of total M equals the equilibrium concentration of free M at the same fixed concentration of total S, the initial velocities are also symmetrical with respect to total S and M. Therefore, the reciprocal plots will be linear whether plotted as a function of free or total S or M. Consequently, it is not possible by this kinetic approach to determine the true substrate for the enzyme. The best procedure is to determine whether each component of the reaction forms complexes with the enzyme or with each other and to determine their dissociation constants by methods discussed earlier. Comparing the dissociation constants obtained from binding studies with those obtained from kinetic studies may allow one to decide the true form of the substrate. For example, equilibrium dialysis studies with creatine kinase showed that, while M_gATP and ATP bind to similar extents, free Mg^{2+}

ions bind only to a slight extent indicating that the formation of the E–Mg complex as an intermediate in the formation of the ternary E–Mg–nucleotide complex is unlikely (138). The true substrate for the reaction is then the magnesium–nucleotide complex. It is important to realize, however, that the enzyme–magnesium complex forms to a slight extent and that its formation may provide a minor pathway for the ternary complex before catalysis occurs.

Calculating the concentration of the free and bound forms of the substrate or enzyme is not always as simple as the above discussion implies. Some reaction mixtures contain several ligands that bind metallocations, each of which may also be affected by pH, making the calculation for the concentration of the free cation rather complex. Yeast and liver pyruvate kinases, for example, are activated by monovalent and divalent cations and by fructose 1,6-diphosphate (136). Because the substrates, adenosine diphosphate (ADP) and phosphoenolpyruvate, the activator, fructose 1,6-diphosphate, and enzyme bind both monovalent and divalent cations, and the monovalent cation and pH affect the binding of divalent cation to the phosphorylated compounds, calculating the concentration of the divalent cation complex and free divalent cation concentration can be complex. Reference (139) however, not only outlines a procedure for calculating the concentration of all species in a mixture of several ionic components that associate but also lists values of dissociation constants for many cation substrate complexes and provides computer programs for completing the calculations.

For enzymes requiring a metal–substrate complex as the active form of the substrate, it is preferable to maintain a constant excess of metal ion over total substrate concentration rather than a constant stoichiometric ratio between the two. Maintaining a constant excess of the metal ion over the substrate ensures that the proportion of the substrate present as the cation complex is maximized and remains constant over a large range of substrate concentrations. Keeping the total concentration of substrate and metal ion cofactor in a constant ratio does not give a constant proportion of the complex (139).

ALLOSTERIC ENZYME KINETICS

One recent important development in the study of metabolism is the discovery that one or more key enzymes in a metabolic pathway exhibit sigmoid kinetics. Sigmoid enzyme kinetics, now called allosteric kinetics, soon became and still is the focus of many investigations because it gives insight into how metabolic pathways can be regulated.

Most allosteric kinetic theory is built on equilibrium ligand binding models that describe the cooperative binding of a ligand to a protein. The model described by Monod et al. (140) supposes a multisubunit enzyme in two different conformations in equilibrium with each other, the R (relaxed) conformation and the T (taut) conformation. In this model, all sites on the R conformer have identical site dissociation constants that are different from those on the T conformer. If the equilibrium between the two conformers favors the form that has a low affinity for the ligand, ligand or substrate will bind preferentially to the form having the highest affinity, and shift the conformer equilibrium toward that form. The model proposed by Monod et al. assumes that all subunits in the protein change conformation simultaneously, that is, they assume symmetry is conserved.

The other popular theory is built on ideas put forth by Adair (141), Pauling (142), and Koshland (143). A distinguishing feature of this model is that each subunit in the allosteric enzyme can assume different conformations in response to binding a ligand; symmetry is not preserved. Therefore, this model has more flexibility and is useful for explaining negative cooperativity, which is not allowed by the model proposed by Monod et al. (140).

Because the two models discussed above are based on equilibrium ligand binding, they often fail when kinetic constraints are imposed. For example, if the conversion of one conformer to another, following binding of a ligand, is slow compared to the rate of the catalyzed reaction, then both models are no longer applicable. In this case, a lag or a transient in the kinetic response curve is observed. Frieden (144) termed this transient enzyme kinetics behavior—hysteretic enzyme kinetics.

Positive and Negative Cooperativity

If the subunits of the dimeric protein molecule depicted in Figure 5.2 are identical, it is reasonable to assume that the dissociation constants of sites 1 and 2 would be identical. Yet there is no chemical or thermodynamic reason that they should be. The primary constraint on the binding reaction is defined by the principle of microscopic reversibility, which requires that the four site constants given in Figure 5.2 are related to one another through the following relationship:

$$K_{D1}K_{D3} = K_{D2}K_{D4} \tag{5.38}$$

This identity forms the basis of the linked function or reciprocity effect. It simply states that the binding of every component of a system is linked with that of every other. If the binding of a ligand at one site does not affect the binding of ligand at another site, the two sites may not be in the same linkage group or, as discussed later, the binding is statistical.

Because the linkage function is an important physical principle, Weber (145) examined the thermodynamics of the process in some detail. Restating Equation (5.38) in terms of free energies gives Equation (5.39), the free energy of conservation

$$\Delta G_1 + \Delta G_3 = \Delta G_2 + \Delta G_4 \tag{5.39}$$

Rearranging Equation (5.39) gives

$$\Delta G_3 - \Delta G_2 = \Delta G_4 - \Delta G_1 = \Delta G_c \tag{5.40}$$

where ΔG_c is defined as the coupling free energy. Assuming $\Delta G_{LPL} = \Delta G_2 + \Delta G_4$ and $\Delta G_c = \Delta G_4 - \Delta G_1$, it follows that

$$\Delta G_c = \Delta G_{LPL} - (\Delta G_1 + \Delta G_2) \tag{5.41}$$

and, therefore, ΔG_c is the difference between the sum of the free energies of formation of PL and LP and the free energy of formation of the biliganded species LPL from the protein and the two ligands.

Weber (145) equates the coupling free energy to the free energy of the following *hypothetical* reaction:

$$\text{PL} + \text{LP} \rightleftharpoons \text{P} + \text{LPL}$$

where

$$K_c = [\text{LPL}][\text{P}]/[\text{PL}][\text{LP}]$$

If $K_{D1} = K_{D2} = K_{D3} = K_{D4}$, $K_c = 1$ and $\Delta G_c = 0$. The concentration of the various enzyme complexes is then equal at equilibrium

$$[LPL] = [P] = [PL] = [LP]$$

In this example, the protein molecule with one site free has the same affinity for the ligand as the molecule with both sites free. If $K_c > 1$, $\Delta G_c < 0$ and LPL and P predominate in the hypothetical equilibrium so that the protein molecule with one site free has a higher affinity for the ligand than the protein molecule with both sites free. This behavior is called positive cooperativity. Negative cooperativity portrays the opposite behavior when $K_c < 1$, so that $\Delta G_c > 0$ and PL and LP predominate in the hypothetical reaction at equilibrium.

The linkage function has important consequences in enzymology. Suppose a protein molecule exists in two conformations and that the transition between these two conformations is associated with the binding of an activator and an inhibitor. It is a necessary consequence of the laws of thermodynamics which define the linkage function that the K_D for binding the activator must be affected by the binding of the inhibitor and vice versa. If the binding of a ligand also alters the subunit interaction forces, the changes may be large enough so that a normally multimeric protein complex may dissociate into its monomeric subunits or a normally monomeric protein may associate into a multimeric complex (146). Changing the concentration of the protein then will not only affect the subunit association but will also affect the K_D for ligand binding.

Because the linear transformation plots used to analyze normal enzyme kinetic data cannot be used to analyze sigmoid kinetic data, most investigators use the Hill plot (1, 3, 4). In 1910, Hill (147) realized that the sigmoid saturation curves of hemoglobin could be described by Equation (5.42) or equations of similar form

$$[PL]/[P_T] = [L]^H/(K_D + [L]^H) \tag{5.42}$$

where H is the Hill slope. Because the Michaelis–Menten equation can be written in a similar form,

$$v/V = [S]^H/(K_m + [S]^H) \tag{5.43}$$

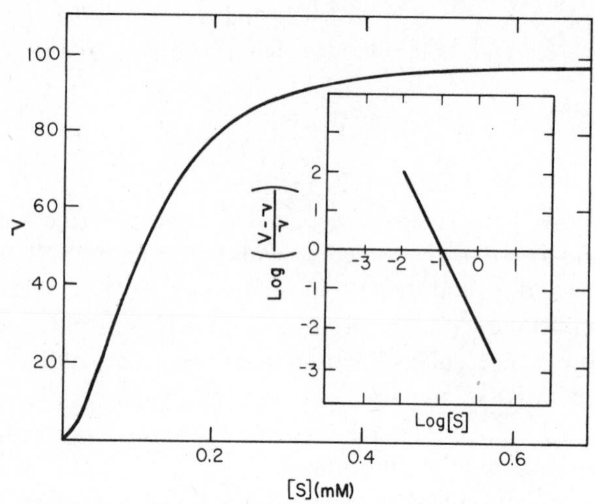

Fig. 5.17. Hypothetic sigmoid enzyme kinetic data. The inset shows the Hill plot of these same sigmoid data as defined by Equation (5.44).

it is also used to treat sigmoid enzyme kinetic data. Rearranging Equation (5.43) and taking the log of both sides gives

$$\log((V - v)/v) = \log K_m - H \log [S] \qquad (5.44)$$

now also known as the Hill equation. A plot of $\log(V - v)/(v)$ versus log [S] gives the Hill plot depicted in Figure 5.17.

The major difficulty with this treatment is that it requires a value of V before the data can be plotted. For a manual fit of the data, the usual procedure is to first estimate V from a v versus [S] plot, then to construct a Hill plot using the estimated V value, and finally to calculate the Hill slope H. A more precise value of H is obtained by first estimating V from a plot of $1/v$ versus $1/[S]^H$, and then constructing a second Hill plot using the best V value. The substrate concentration at 0.5 V is the apparent K_m value called $[S]_{0.5}$. A still more precise and objective method for analyzing sigmoid kinetic data requires a nonlinear curve fitting computer program (113, 115, 118).

At this point in the discussion, it is important to emphasize that some sigmoid enzyme kinetic data are more complex than the above treatment implies. For example, initial velocities of AMP aminohydrolase plotted as a function of AMP concentrations at pH 7.0 were sigmoid when enzyme concentrations was 0.5 μg/mL but hyperbolic at 50 μg/mL. At pH 6.3, on the other hand, profiles of substrate saturation versus initial velocities were hyperbolic at both protein concentrations (148). The data are consistent with a protein concentration dependent, dilution induced inactivation of the enzyme, which can be influenced by H^+ and substrate concentration. At pH 7.0 and at low protein and substrate concentrations, the enzyme dissociates into inactive subunits. The initial velocity of the enzyme catalyzed reaction is then lower than expected because the rate reflects both loss of enzyme activity and the catalytic reaction. If the multimeric enzyme containing bound substrate does not dissociate into its subunits at pH 7.0, then the expected catalytic rates will be observed at high substrate concentrations and a plot of initial velocities versus substrate concentration will now be sigmoid. Therefore, if dilution inactivation of the enzyme is suspected as an explanation for the sigmoid kinetic behavior, complete the reaction at higher enzyme concentrations, using a stopped flow device if necessary, as a test.

If the dilution inactivation of the protein is reversible, then the role of the protein concentration on the allosteric kinetic system can also be studied. The oligomerization reaction, however, need not be a linked function as observed with *E. coli* threonine dehydrase (149). Adenosine monophosphate activates this enzyme by causing a decrease in the K_m for threonine and an increase in the affinities between the catalytically active subunits of the dimeric protein. Yet, oligomers formed at high protein concentration (1 mg/mL), in the absence of the activator AMP, also have a high K_m for the substrate. Therefore, oligomer formation alone is not a sufficient condition to decrease the K_m for threonine. The data indicate that the oligomerization reaction is not linked to the decrease in K_m for substrate. On the other hand, numerous multisubunit proteins show a linkage between the oligomerization reaction and the binding of ligands. Hemoglobin provides one such example (146). At neutral pH, the conformative response of hemoglobin to oxygenation is a new conformer that has weaker forces of attraction between the two α–β dimers. Oxygenation of hemoglobin then, depending on its total concentration, promotes a reversible dissociation into its α–β dimers.

REFERENCES

1. A. Cornish-Bowden, *Fundamentals of Enzyme Kinetics*, Butterworths, London, 1979.
2. A. Cornish-Bowden, *Biochem. J.*, **153**, 455–461 (1976).

3. I. H. Segel, *Enzyme Kinetics*, Wiley, New York, 1975.
4. S. Ainsworth, *Steady-State Enzyme Kinetics*, University Park, Baltimore, MD, 1977
5. M. Dixon and E. C. Webb, *Enzymes*, 3rd ed., Academic, New York, 1979, pp. 369–381.
6. C. H. Suelter and W. Melander, *J. Biol. Chem.,* **238,** PC4108–PC4109 (1963).
7. E. G. Gorman and D. W. Darnall, *Biochemistry,* **20,** 38–43 (1981).
8. I. Feldman and D. C. Kramp, *Biochemistry,* **17,** 1541–1547 (1978).
9. N. J. Greenfield, M. N. Williams, M. Poe, and K. Hoogsteen, *Biochemistry,* **11,** 4706–4711 (1972).
10. I. M. Chaiken, *Anal. Biochem.,* **97,** 1–10 (1979).
11. I. M. Chaiken and H. C. Taylor, *J. Biol. Chem.,* **251,** 2044–2048 (1976).
12. B. M. Dunn, and I. M. Chaiken, *Biochemistry,* **14,** 2343–2349 (1975).
13. K. Takeo and S. Nakamura, *Arch. Biochem. Biophys.,* **153,** 1–7 (1972).
14. V. Horejsi, *J. Chromatogr.,* **178,** 1–13 (1979).
15. S. J. Johnson, E. C. Metcalf, and P. D. G. Dean, *Anal. Biochem.,* **109,** 63–66 (1980).
16. V. Horejsi, M. Ticha, P. Tichy, and A. Holy, *Anal. Biochem.,* **125,** 358–369 (1982).
17. I. M. Klotz, *Acc. Chem. Res.,* **7,** 162–168 (1974).
18. I. M. Klotz, F. M. Walker, and R. B. Pivan, *J. Am. Chem. Soc.,* **68,** 1486–1490 (1946).
19. R. H. Reitz and W. H. Riley, *Anal. Biochem.,* **36,** 535–537 (1970).
20. C. E. Furlong, R. G. Morris, M. Kandrach, and B. P. Rosen, *Anal. Biochem.,* **47,** 514–526 (1972).
21. P. T. Englund, J. A. Huberman, T. M. Jovin, and A. Kornberg, *J. Biol. chem.,* **244,** 3038–3044 (1969).
22. V. P. Demushkin and N. M. Shalina, *Anal. Biochem.,* **105,** 230–232 (1980).
23. C. Tanford, *Physical Chemistry of Macromolecules*, Wiley, New York, 1961, pp. 225–227.
24. P. Suter and J. P. Rosenbusch, *Anal. Biochem.,* **82,** 109–114 (1977).
25. J. A. Sophianopoulos, S. J. Durham, A. J. Sophianopoulos, H. L. Ragsdale, and W. P. Cropper, Jr., *Arch. Biochem. Biophys.,* **187,** 132–137 (1978).
26. J. T. Edsall and J. Wyman, *Biophysical Chemistry*, Vol. 1, Academic, New York, 1958.
27. H. Paulus, *Anal. Biochem.,* **32,** 91–100 (1969).
28. P. Suter and J. P. Rosenbusch, *J. Biol. Chem.,* **251,** 5986–5991 (1976).
29. P. A. Lazo, A. Sols, and J. E. Wilson, *J. Biol. Chem.,* **255,** 7548–7551 (1980).
30. R. Freundlich and D. B. Taylor, *Anal. Biochem.,* **114,** 103–104 (1981).
31. C. E. Spivak and D. B. Taylor, *Anal. Biochem.,* **77,** 274–279 (1977).
32. E. Heyde, *Anal. Biochem.,* **51,** 61–66 (1973).
33. S. P. Colowick and F. C. Womack, *J. Biol. Chem.,* **244,** 774–777 (1969).
34. F. C. Womack and S. P. Colowick, *Methods Enzymol.,* **27,** 464–471 (1973).
35. K. Feldmann, *Anal. Biochem.,* **88,** 225–235 (1978).
36. L. C. Craig and T. P. King, *Methods Biochem. Anal.,* **10,** 175–199 (1962).
37. J. P. M. Wielders, *Anal. Biochem.,* **103,** 394–399 (1980).
38. G. K. Ackers, *Methods Enzymol.,* **27,** 441–464 (1973).

39. J. P. Hummel and W. J. Dreyer, *Biochim. Biophys. Acta,* **63,** 530–532 (1962).
40. H. S. Penefsky, *J. Biol. Chem.,* **252,** 2891–2899 (1977).
41. K. B. Andersen and M. H. Vaughan, *J. Chromatogr.,* **240,** 1–8 (1982).
42. B. Zeeberg and M. Caplow, *Biochemistry,* **18,** 3880–3886 (1979).
43. S. F. Velick, J. E. Hayes, Jr., and J. Harting, *J. Biol. Chem.,* **203,** 527–544 (1953).
44. H. Hirsch-Kolb, H. J. Kolb, and D. M. Greenberg, *Anal. Biochem.,* **34,** 517–528 (1970).
45. N. J. Greenfield, *CRC Crit. Rev. Biochem.,* **3,** 71–110 (1975).
46. K. Zierler, *Biophys. Struct. Mech.,* **3,** 275–289 (1977).
47. K. C. Ingham and C. H. Suelter, *Arch. Biochem. Biophys.,* **168,** 531–535 (1975).
48. A. Stockell, *J. Biol. Chem.,* **234,** 1286–1292 (1959).
49. L. Stryer, *J. Mol. Biol.,* **13,** 482–495 (1965).
50. S. E. Builder, J. A. Beavo, and E. G. Krebs, *J. Biol. Chem.,* **255,** 2350–2354 (1980).
51. M. Epstein, A. Levitzki, and J. Reuben, *Biochemistry,* **13,** 1777–1782 (1974).
52. A. S. Mildvan and J. L. Engle, *Methods Enzymol.,* **26C,** 654–682 (1972).
53. A. S. Mildvan and M. Cohn, *Biochemistry,* **2,** 910–919 (1963).
54. K. M. Valentine and G. L. Cottam, *Arch. Biochem. Biophys.,* **158,** 346–354 (1973).
55. C. J. Halfman and T. Nishida, *Biochemistry,* **11,** 3493–3498 (1972).
56. M. P. Heyn and W. O. Weischet, *Biochemistry,* **14,** 2962–2968 (1975).
57. P. Devaux, R. Viennet, and M. Legrand, *FEBS Lett.,* **40,** 18–24 (1974).
58. T. Kodama and R. C. Woledge, *J. Biol. Chem.,* **251,** 7499–7503 (1976).
59. P. Coassolo, M. Sarrazin, and J. C. Sari, *Anal. Biochem.,* **104,** 37–43 (1980).
60. G. Kegeles, *Methods Enzymol.,* **27,** 456–464 (1973).
61. C. W. Gray and M. J. Chamberlin, *Anal. Biochem.,* **41,** 83–104 (1971).
62. J. Wallach and M. Hanss, *Anal. Biochem.,* **88,** 69–77 (1978).
63. G. A. Rechnitz, *Science,* **214,** 287–291 (1981).
64. P. W. Carr and L. D. Bowers, *Theory and Applications of Enzyme Electrodes in Immobilized Enzymes in Analytical and Clinical Chemistry*, Chap. 5, Wiley, New York, 1980, pp. 197–310.
65. M. D. Hollenberg and P. Cuatrecasas, Distinction of receptor from nonreceptor interactions in binding studies, in R. D. O'Brien, Ed., *The Receptors*, Vol. 1, Plenum, New York, 1979, pp. 193–214.
66. D. Colquhoun, The link between drug binding and response: Theories and observations, in R. D. O'Brien Ed., *The Receptors*, Vol. 1, Plenum, New York, 1979, pp. 93–142.
67. Y-C. Cheng and W. H. Prusoff, *Biochem. Pharmacol.,* **22,** 3099–3108 (1973).
68. K-J. Chang, S. Jacobs, and P. Cuatrecasas, *Biochim. Biophys. Acta,* **406,** 294–303 (1975).
69. R. D. O'Brien, Problems and approaches in noncatalytic biochemistry, in R. D. O'-Brien, Ed., *The Receptors*, Vol. 1, Plenum, New York, 1979, pp. 311–335.
70. K. C. Ingham, *Anal. Biochem.,* **68,** 660–663 (1975).
71. C. Y. Huang, *Methods Enzymol.,* **87,** 509–525 (1982).
72. D. E. Koshland, Jr., in P. D. Boyer, H. Lardy, and K. Myrback, Eds., *The Enzymes*, Vol. 1, Academic, New York, 1959, pp. 305–346.

73. N. Citri, *Adv. Enzymol. Relat. Areas Mol. Biol.,* **37,** 397–648 (1973).
74. W. N. Lipscomb, *Ann. Rev. Biochem.,* **52,** 17–34 (1983).
75. F. J. Kayne and C. H. Suelter, *J. Am. Chem. Soc.,* **87,** 897–900 (1965).
76. C. H. Suelter, *Biochemistry,* **6,** 418–423 (1967).
77. A. S. Mildvan, *Acc. Chem. Res.,* **10,** 246–252 (1977).
78. L. Stryer, *Science,* **162,** 526–533 (1968).
79. G. M. Edelman and W. O. McClure, *Acc. Chem. Res.,* **1,** 65–70 (1968).
80. R. Jones, R. A. Dwek, and I. O. Walker, *FEBS Lett.,* **26,** 92–96 (1972).
81. J-R. Garel, *Eur. J. Biochem.,* **70,** 179–189 (1976).
82. S. Yun and C. H. Suelter, *J. Biol. Chem.,* **254,** 1806–1810 (1979).
83. S. W. Englander and N. R. Kallenback, *Q. Rev. Biophys.,* **16,** 521–655 (1983).
84. K. Pfister, J. H. R. Kagi, and P. Christen, *Proc. Nat. Acad. Sci. USA,* **75,** 145–148 (1978).
85. S. W. Englander and J. J. Englander, *Methods Enzymol.,* **26,** 406–413 (1972).
86. E. L. Malin and S. W. Englander, *J. Biol. Chem.,* **255,** 10,695–10,701 (1980).
87. R. K. H. Liem, D. B. Calhoun, J. J. Englander, and S. W. Englander, *J. Biol. Chem.,* **255,** 10,687–10,694 (1980).
88. F. J. Kayne and C. H. Suelter, *Biochemistry,* **7,** 1678–1684 (1968).
89. C. Frieden, *J. Biol. Chem.,* **245,** 5788–5799 (1970).
90. B. Furie, A. N. Schechter, D. H. Sachs, and C. B. Anfinsen, *J. Mol. Biol.,* **92,** 497–506 (1975).
91. J. G. R. Hurrell, J. A. Smith, and S. J. Leach, *Biochemistry,* **16,** 175–185 (1977).
92. M. W. Kirschner and H. K. Schachman, *Biochemistry,* **10,** 1900–1919 (1971).
93. R. D. Allison and D. L. Purich, *Methods Enzymol.,* **63,** 3–23 (1979).
94. K. F. Tipton, Enzyme assay and kinetic studies, in *Techniques in Protein and Enzyme Biochemistry*, Part II, B112, Elsevier–North Holland, Amsterdam, 1978, pp. 1–56.
95. G. L. Atkins and I. A. Nimmo, *Anal. Biochem.,* **104,** 1–9 (1980).
96. H. J. Lee and I. B. Wilson, *Biochim. Biophys. Acta,* **242,** 519–522 (1971).
97. S. G. Waley, *Biochem. J.,* **193,** 1009–1012 (1981).
98. A. Cornish-Bowden, *Biochem. J.,* **149,** 305–312 (1975).
99. D. A. Deranleau, *J. Am. Chem. Soc.,* **91,** 4044–4049 (1969).
100. R. C. Kohberger, *Anal. Biochem.,* **101,** 1–6 (1980).
101. T. M. Wonnacott and R. J. Wonnacott, *Introductory Statistics*, Wiley, New York, 1969, pp. 136–137.
102. B. Mannervik, *Methods Enzymol.,* **87,** 370–390 (1982).
103. W. W. Cleland, *Adv. Enzymol. Relat. Areas Mol. Biol.,* **29,** 1–32 (1967).
104. P. J. F. Henderson, Statistical analysis of enzyme kinetic data, in *Techniques in Protein and Enzyme Biochemistry*, B113, Elsevier–North Holland, Amsterdam, 1978, pp. 1–43.
105. R. G. Duggleby, Experimental designs for the distribution-free analysis of enzyme kinetic data, in L. Endrenyi, Ed., *Kinetic Data Analysis: Design and Analysis of Enzyme and Parmacokinetic Experiments*, Plenum, New York, 1981, pp. 169–179.
106. D. Burk, *Trends Biochem. Sci.,* **9,** 202–204 (1984).
107. I. M. Klotz and D. L. Hunston, *Arch. Biochem. Biophys.,* **193,** 314–328 (1979).

108. S. Ferguson-Miller and W. H. Koppenol. *Trends Biochem. Sci.*, **6**, 4–7 (1981).

109. B. Mannervik, Design and analysis of kinetic experiments for discrimination between rival models, in L. Endrenyi, Ed., *Kinetic Data Analysis. Design and Analysis of Enzyme and Pharmacokinetic Experiments*, Plenum, New York, 1981, pp. 235–270.

110. W. W. Cleland, *Methods Enzymol.*, **63**, 103–138 (1979).

111. G. N. Wilkinson, *Biochem. J.*, **80**, 324–332 (1961).

112. C. H. Suelter, A. J. Morris, and D. Hill, *J. Chem. Ed.*, **58**, 988–989 (1981).

113. J. L. Dye and V. A. Nicely, *J. Chem. Ed.*, **48**, 443–448 (1971).

114. G. L. Atkins, *Biochim. Biophys. Acta,* **252**, 421–426 (1971).

115. I. Knack and K-H. Rohm, *Hoppe-Seyler's Z. Physiol. Chem.*, **362**, 1119–1130 (1981).

116. R. D. Duggleby, *Anal. Biochem.*, **110**, 9–18 (1981).

117. R. Bianchi, G. M. Hanozet, and M. P. Simonetta, *Comput. Programs Biomed.*, **16**, 189–194 (1983).

118. M. J. C. Crabbe, *Comput. Biol. Med.*, **12**, 263–283 (1982).

119. E. I. Canela, *Int. J. Bio-Med. Comput.*, **15**, 121–130 (1984).

120. L. Garfinkel, M. C. Kohn, and D. Garfinkel, *CRC. Crit. Rev. Bioeng.* **2**, 329–361 (1977).

121. D. I. Little, P. C. Poat, and I. G. Giles, *Eur. J. Biochem.*, **124**, 499–505 (1982).

122. B. Mannervik, I Jakobson, and M. Warholm, *Biochim. Biophys. Acta,* **567**, 43–48 (1979).

123. L. A. Schriefer and W. F. Benisek, *Anal. Biochem.*, **141**, 437–445 (1984).

124. B. A. Orsi and K. F. Tipton, *Methods Enzymol.*, **63**, 159–183 (1979).

125. S. Yun and C. H. Suelter, *Biochim. Biophys. Acta,* **480**, 1–13 (1977).

126. Y. Fukagawa, *Biochem. J.*, **185**, 186–188 (1980).

127. D. J. Bates and C. Frieden, *Comput. Biomed. Res.*, **6**, 474–486 (1973).

128. D. B. Northrop, *Anal. Biochem.*, **132**, 457–461 (1983).

129. W. W. Cleland, *Acc. Chem. Res.* **8**, 145–151 (1975).

130. N. C. Price, *Trends Biochem. Sci.*, **4**, N272–N273 (1979).

131. C. N. Pace, *Trends Biochem. Sci.*, **5**, 173–174 (1980).

132. S. Cha, *Biochem. Pharmacol.*, **24**, 2177–2185 (1975).

133. T. Alston, D. Porter, and H. Bright, *Acc. Chem. Res.*, **16**, 418–424 (1983).

134. J. F. Morrison, *Trends Biochem. Sci.*, **7**, 102–106 (1982).

135. C. T. Walsh, *Ann. Rev. Biochem.*, **53**, 493–535 (1984).

136. T. Nowak and C. Suelter, *Mol. Cell. Biochem.*, **35**, 65–75 (1981).

137. Y. H. Baek and T. Nowak, *Arch. Biochem. Biophys.*, **217**, 491–497 (1982).

138. D. C. Watts, Creatine kinase (adenosine-5′-triphosphate-creatine phosphotransferase), in P. D. Boyer, Ed., *The Enzymes*, 3rd ed., Academic, New York, Vol. 8, 1973, pp. 383–455.

139. W. J. O'Sullivan and G. W. Smithers, *Methods Enzymol.*, **63**, 294–336 (1979).

140. J. Monod, J. Wyman, and J.-P. Changeux, *J. Mol. Biol.*, **12**, 88–118 (1965).

141. G. S. Adair, *J. Biol. Chem.*, **63**, 529–545 (1925).

142. L. Pauling, *Proc. Nat. Acad. Sci. USA,* **21**, 186–191 (1935).

143. D. E. Koshland, Jr., The molecular basis for enzyme regulation, in P. D. Boyer, Ed., *The Enzymes*, 3rd. ed., Academic, New York, Vol. 1, 1970, pp. 341–396.

144. C. Frieden, *J. Biol. Chem.*, **245**, 5788–5799 (1970).
145. G. Weber, *Adv. Protein Chem.*, **29**, 1–83 (1975).
146. B. W. Turner, D. W. Pettigrew, and G. K. Ackers, *Methods Enzymol.*, **76**, 596–628 (1981).
147. A. V. Hill, *J. Physiol. (London)*, **40**, 4–7 (1910).
148. R. M. Hemphill, C. L. Zielke, and C. H. Suelter, *J. Biol. Chem.*, **246**, 7237–7240 (1971).
149. C. P. Dunne and W. A. Wood, *Curr. Top. Cell. Regul.*, **9**, 65–101 (1975).

APPENDIX I

Composition of Stock Solutions Used in Polyacrylamide Gel Electrophoresis

I. STOCK SOLUTIONS FOR RUNNING GEL*

IA: Acrylamide–bisacrylamide solution

IA1: Acrylamide–glycerol solution

Acrylamide	39.0 g
N,N'-methylenebisacrylamide	1.0 g
Glycerol	20.0 mL
H_2O to make 100 mL	

IA2: Acrylamide

Acrylamide	39.0 g
N,N'-methylenebisacrylamide	1.0 g
H_2O to make 100 mL	

IB: Buffer solutions†

IB1: Tris–HCl buffer (0.75 *M* Tris)

Tris	9.15 g
HCl (conc)	3.0 mL
H_2O to make 100 mL	

* Abstracted with modification from Z. I. Ogita and C. L. Markert, *Anal. Biochem.* **99,** 233–241 (1979).

† The composition of other buffers ranging in pH from 3.8 to 10.2 is given later in Appendix I.

IB2: Tris–HCl–Triton buffer (3.0 *M* Tris)

Tris	36.6 g
HCl (conc)	12.0 mL
20% Triton X-100	20.0 mL
H_2O to make 100 mL	

IB3: Tris–HCl–SDS buffer (1.5 *M* Tris)

Tris	18.3 g
HCl (conc)	6.0 mL
10% Sodium dodecyl sulfate in H_2O	4.0 mL
H_2O to make 100 mL	

IB4: Phosphate–SDS buffer (0.1 *M*)

$Na_2HPO_4 \cdot 7H_2O$	3.86 g
$NaH_2PO_4 \cdot H_2O$	0.78 g
10% Sodium dodecyl sulfate in H_2O	4.0 mL
H_2O to make 100 mL	

IC: 0.2 % (w/v) APS solution

Ammonium persulfate	0.2 g
H_2O to make 100 mL	

ID: 0.4% (v/v) Temed solution

N,N,N′N′-tetramethylenediamine	0.4 mL
H_2O to make 100 mL	

II. STOCK SOLUTIONS FOR STACKING GEL*

IIA: Acrylamide–bisacrylamide solution

IIA1: Acrylamide–bisacrylamide–glycerol

Acrylamide	38.0 g
N,N′-methylenebisacrylamide	2.0 g
Glycerol	20.0 mL
H_2O to make 100 mL	

IIA2: Acrylamide–bisacrylamide

Acrylamide	38.0 g
N,N′-methylenebisacrylamide	2.0 g
H_2O to make 100 mL	

* Abstracted with modification from Z. I. Ogita and C. L. Markert, *Anal. Biochem.* **99**, 233–241 (1979).

IIB: Buffer solutions

IIB1: Tris–HCl buffer (0.125 *M* Tris)

Tris	1.5 g
HCl (conc)	1.0 mL
H_2O to make 100 mL	

IIB2: Tris–HCl–Triton buffer (0.50 *M* Tris)

Tris	6.0 g
HCl (conc)	4.0 mL
20% Triton X-100 in H_2O	20.0 mL
H_2O to make 100 mL	

IIB3: Tris–HCl–SDS buffer (0.25 *M* Tris)

Tris	3.0 g
HCl (conc)	2.0 mL
10% Sodium dodecyl sulfate in H_2O	4.0 mL
H_2O to make 100 mL	

IIB4: Phosphate–SDS buffer (0.025 *M*, pH 7.2)

See IB4	25.0 mL
H_2O to make 100 mL	

IIC: 0.2% (w/v) APS solution

Ammonium persulfate	0.4 g
H_2O to make 100 mL	

IID: 2.0% (v/v) Temed solution

N,N,N'N'-Tetramethylenediamine	2.0 mL
H_2O to make 100 mL	

III. ELECTRODE BUFFERS†

IIIA: Tris–glycine buffer (0.0125 *M* Tris) pH 8.3

Tris	1.5 g
Glycine	7.2 g
H_2O to make 1.0 L	

IIIB: Tris–glycine–Triton buffer (0.05 *M* Tris) pH 8.3

Tris	6.0 g
Glycine	28.8g
20% Triton X-100 in H_2O	50.0 mL

† The composition of other buffers ranging in pH from 3.8 to 10.2 is given later in Appendix I.

	H_2O to make 1.0 L	
IIIC:	Tris–glycine–SDS buffer (0.025 *M* Tris) pH 8.3	
	Tris	3.0g
	Glycine	14.4 g
	10% Sodium dodecyl sulfare in H_2O	10.0 mL
	H_2O to make 1.0 L	
IIID:	Phosphate–SDS buffer (0.025 *M*) pH 7.2	
	see IB4	25.0 mL
	H_2O to make 100 mL	

IV. STOCK SOLUTIONS FOR SAMPLE PREPARATION*

IVA:	Sample diluting solution (without detergents)	
	0.5 *M* Tris–HCl buffer (pH 6.8)	25.0 mL
	Glycerol	40.0 mL
	0.01% Bromophenol blue	20.0 mL
	H_2O	15.0 mL
IVB:	Sample diluting solution with Triton X-100	
	0.5 *M* Tris–HCl buffer (pH 6.8)	25.0 mL
	20% Triton X-100 in H_2O	10.0 mL
	Glycerol	40.0 mL
	0.01% Pyronine Y	20.0 mL
	H_2O	5.0 mL
IVC:	2.5% SDS solution for treating samples	
	0.5 *M* Tris–HCl buffer (pH 6.8)	3.0 mL
	10% Sodium dodecyl sulfate in H_2O	5.0 mL
	Glycerol	10.0 mL
	2-mercaptoethanol	2.0 mL
IVD:	Diluting solution (after SDS treatment)	
	0.5 *M* Tris–HCl buffer (pH 6.8)	1.5 mL
	10% Sodium dodecyl sulfate in H_2O	2.5 mL
	Glycerol	5.0 mL
	0.005% Bromophenol blue	4.0 mL
	H_2O	7.0 mL

* Abstracted with modification from Z. I. Ogita and C. L. Markert, *Anal. Biochem.* **99**, 233–241 (1979).

IVE: Bromophenol blue solution (0.005%)

1% Bromophenol blue in methanol	0.5 mL
H_2O	100.0 mL

V. COMPOSITION OF POLYACRYLAMIDE GELS OF DIFFERENT CONCENTRATIONS*,†

Composition of running gels			Composition of stacking gels		
	Stock Solution			Stock Solution	
Percentage Gel	IA (mL)	H_2O (mL)	Percentage Gel	IIA (mL)	H_2O (mL)
3	1.5	8.5	3.0	1.50	8.50
4	2.0	8.0	3.5	1.75	8.25
5	2.5	7.5	4.0	2.00	8.00
6	3.0	7.0	4.5	2.25	7.75
7	3.5	6.5	5.0	2.50	7.50
8	4.0	6.0	5.5	2.75	7.25
9	4.5	5.5	6.0	3.00	7.00
10	5.0	5.0			
11	5.5	4.5			
12	6.0	4.0			
13	6.5	3.5			
14	7.0	3.0			
15	7.5	2.5			
16	8.0	2.0			
17	8.5	1.5			
18	9.0	1.0			
19	9.5	0.5			
20	10.0	0			

* Abstracted with permission from Z. I. Ogita and C. L. Markert, *Anal. Biochem.*, **99**, 233–241 (1979).

† In preparing each concentration of the gel according to this table, also add 5 mL of solution B, 2.5 mL of solution C, and 2.5 mL of solution D. For running gels, add the solutions in list I and for stacking gels, add the solutions defined in list II.

VI COMPOSITION AND CHARACTERIZATION OF ELECTROPHORESIS BUFFERS[a]

			After 8 hr at 300 V[b]			Buffering Capacity[c]	
			pH				
pH	Basic component	Acidic Component	Top	Bottom	Change in pH	Acid	Base
3.8	30 m*M* β-alanine	20 m*M* lactic acid	4.36	3.79	0.43	2.4	1.8
4.4	80 m*M* β-alanine	40 m*M* acetic acid	4.49	4.33	0.16	4.9	1.9
4.8	80 m*M* Gaba	20 m*M* acetic acid	5.02	4.74	0.28	4.9	1.6
6.1	30 m*M* histidine	30 m*M* Mes[d]	6.11	6.05	0.06	2.4	2.1
6.6	25 m*M* histidine	30 m*M* Mops	6.64	6.52	0.12	1.9	1.9
7.4	43 m*M* imidazole	35 m*M* Hepes	7.59	7.29	0.30	3.3	2.5
8.1	32 m*M* Tris	30 m*M* Epps	8.58	8.06	0.52	2.4	1.2
8.7	50 m*M* Tris	25 m*M* boric acid	8.81	8.69	0.12	2.9	1.6
9.4	60 m*M* Tris	40 m*M* Caps	9.43	9.36	0.07	2.2	1.4
10.2	37 m*M* ammonia	20 m*M* Caps	10.37	10.19	0.18	2.7	1.5

[a] Abstracted with permission from T. McLellan, *Anal. Biochem.*, **126**, 94–99 (1982).
[b] pH of electrode compartments; top is the cathode, bottom the anode. Change in pH is the difference between them.
[c] Volume, in milliliters, of 1 *N* HCl or NaOH added to 100 mL of buffer to change pH by 1 unit.
[d] Abbreviations used: Caps, 3-(cyclohexylamino)-1-propanesulfonic acid; Epps, *N*-(2-hydroxyethyl)piperazine-*N'*-3-propanesulfonic acid; Gaba, γ-amino-*n*-butyric acid; Hepes, *N*-2-hydroxyethylpiperazine-*N'*-2-ethanesulfonic acid; Mes, 2-(*N*-morpholino)-ethanesulfonic acid; Mops, 3-(*N*-morpholino)propanesulfonic acid; Tris, tris(hydroxymethyl) aminomethane.

APPENDIX II

The Error Function as Applied to Gel Chromatography Data[a,b]

	$(1 - K_p)$									
$\mathrm{erf}^{-1}(1 - K_p)$	0	1	2	3	4	5	6	7	8	9
0.00	0.0000	0011	0023	0034	0045	0056	0068	0079	0090	0102
0.01	0.0113	0124	0135	0147	0158	0179	0181	0192	0203	0214
0.02	0.0226	0237	0248	0260	0271	0282	0293	0305	0316	0327
0.03	0.0338	0350	0361	0372	0384	0395	0406	0417	0429	0440
0.04	0.0451	0462	0474	0485	0496	0507	0519	0530	0541	0553
0.05	0.0564	0575	0586	0598	0609	0620	0631	0643	0654	0665
0.06	0.0676	0688	0699	0710	0721	0732	0744	0755	0766	0777
0.07	0.0789	0800	0811	0822	0834	0845	0856	0867	0878	0890
0.08	0.0901	0912	0923	0934	0946	0957	0968	0979	0990	1002
0.09	0.1013	1024	1035	1046	1058	1069	1080	1091	1102	1114
0.10	0.1125	1136	1147	1158	1169	1181	1192	1203	1214	1225
0.11	0.1236	1247	1259	1270	1281	1292	1303	1314	1325	1337
0.12	0.1348	1359	1370	1381	1392	1403	1414	1425	1437	1448
0.13	0.1459	1470	1481	1492	1503	1514	1525	1536	1547	1558
0.14	0.1570	1581	1592	1603	1614	1625	1636	1647	1658	1669
0.15	0.1680	1691	1702	1713	1724	1735	1746	1757	1768	1779
0.16	0.1790	1801	1812	1823	1834	1845	1856	1867	1878	1889
0.17	0.1900	1911	1922	1933	1944	1955	1966	1977	1988	1998
0.18	0.2009	2020	2031	2042	2053	2064	2075	2086	2097	2108
0.19	0.2118	2129	2140	2151	2162	2173	2184	2195	2205	2216
0.20	0.2227	2238	2249	2260	2270	2281	2292	2303	2314	2324
0.21	0.2335	2346	2357	2368	2378	2389	2400	2411	2421	2432
0.22	0.2443	2454	2464	2475	2486	2497	2507	2518	2529	2540
0.23	0.2550	2561	2572	2582	2593	2604	2614	2625	2636	2646
0.24	0.2657	2668	2678	2689	2700	2710	2721	2731	2742	2753
0.25	0.2763	2774	2785	2795	2806	2816	2827	2837	2848	2858

	$(1 - K_p)$									
$\mathrm{erf}^{-1}(1 - K_p)$	0	1	2	3	4	5	6	7	8	9
0.26	0.2869	2880	2890	2901	2911	2922	2932	2943	2953	2964
0.27	0.2974	2985	2995	3006	3016	3027	3037	3048	3058	3068
0.28	0.3079	3089	3100	3110	3121	3131	3141	3152	3162	3173
0.29	0.3183	3192	3204	3214	3224	3235	3245	3255	3266	3276
0.30	0.3286	3297	3307	3317	3328	3338	3348	3358	3369	3379
0.31	0.3389	3399	3410	3420	3430	3440	3451	3461	3471	3481
0.32	0.3491	3501	3512	3522	3532	3542	3552	3562	3573	3583
0.33	0.3593	3603	3613	3623	3633	3643	3653	3664	3674	3684
0.34	0.3694	3704	3714	3724	3734	3744	3754	3764	3774	3784
0.35	0.3794	3804	3814	3824	3834	3844	3854	3864	3874	3883
0.36	0.3893	3903	3913	3923	3933	3943	3953	3963	3972	3982
0.37	0.3992	4002	4012	4022	4031	4041	4051	4061	4071	4080
0.38	0.4090	4100	4110	4119	4129	4139	4149	4158	4168	4178
0.39	0.4187	4197	4207	4216	4226	4236	4245	4255	4265	4274
0.40	0.4284	4294	4303	4313	4322	4332	4342	4351	4361	4370
0.41	0.4380	4389	4399	4408	4418	4427	4437	4446	4456	4465
0.42	0.4475	4484	4494	4503	4512	4522	4531	4541	4550	4560
0.43	0.4569	4578	4588	4597	4606	4616	4625	4634	4644	4653
0.44	0.4662	4672	4681	4690	4699	4709	4718	4727	4736	4746
0.45	0.4755	4764	4773	4782	4792	4801	4810	4819	4828	4837
0.46	0.4847	4856	4865	4874	4883	4892	4901	4910	4919	4928
0.47	0.4938	4947	4956	4965	4974	4983	4992	5001	5010	5019
0.48	0.5028	5037	5045	5054	5063	5072	5081	5090	5099	5108
0.49	0.5117	5126	5134	5143	5152	5161	5170	5179	5187	5196
0.50	0.5205	5214	5223	5231	5240	5249	5258	5266	5275	5284
0.51	0.5292	5301	5310	5319	5327	5336	5345	5353	5362	5370
0.52	0.5379	5388	5396	5405	5413	5422	5431	5439	5448	5456
0.53	0.5465	5473	5482	5490	5499	5507	5516	5524	5533	5541
0.54	0.5549	5558	5566	5575	5583	5591	5600	5608	5617	5625
0.55	0.5633	5642	5650	5658	5667	5675	5683	5691	5700	5708
0.56	0.5716	5724	5733	5741	5749	5757	5766	5774	5782	5790
0.57	0.5798	5806	5814	5823	5831	5839	5847	5855	5863	5871
0.58	0.5879	5887	5895	5903	5911	5919	5927	5935	5943	5951
0.59	0.5959	5967	5975	5983	5991	5999	6007	6015	6023	6031
0.60	0.6039	6046	6054	6062	6070	6078	6086	6093	6101	6109
0.61	0.6117	6125	6132	6140	6148	6156	6163	6171	6179	6186
0.62	0.6194	6202	6210	6217	6225	6232	6240	6248	6255	6263
0.63	0.6271	6278	6286	6293	6301	6308	6316	6323	6331	6338
0.64	0.6346	6353	6361	6368	6376	6383	6391	6398	6406	6413
0.65	0.6420	6428	6435	6442	6450	6457	6465	6472	6479	6487

	$(1 - K_p)$									
$\text{erf}^{-1}(1 - K_p)$	0	1	2	3	4	5	6	7	8	9
0.66	0.6494	6501	6508	6516	6523	6530	6537	6545	6552	6559
0.67	0.6566	6574	6581	6588	6595	6602	6609	6617	6624	6631
0.68	0.6638	6645	6652	6659	6666	6673	6680	6687	6694	6701
0.69	0.6708	6715	6722	6729	6736	6743	6750	6757	6764	6771
0.70	0.6778	6785	6792	6799	6806	6813	6819	6826	6833	6840
0.71	0.6847	6854	6860	6867	6874	6881	6887	6894	6901	6908
0.72	0.6915	6921	6928	6934	6941	6948	6955	6961	6968	6974
0.73	0.6981	6988	6994	7001	7008	7014	7021	7027	7034	7040
0.74	0.7047	7053	7060	7066	7073	7079	7086	7092	7099	7105
0.75	0.7112	7118	7124	7131	7137	7144	7150	7156	7163	7169
0.76	0.7175	7182	7188	7194	7201	7207	7213	7220	7226	7232
0.77	0.7238	7244	7251	7257	7263	7269	7276	7282	7288	7294
0.78	0.7300	7306	7312	7319	7325	7331	7337	7343	7349	7355
0.79	0.7361	7367	7373	7379	7385	7391	7397	7403	7409	7415
0.80	0.7421	7427	7433	7439	7445	7451	7457	7462	7468	7474
0.81	0.7480	7486	7492	7498	7503	7509	7515	7521	7527	7532
0.82	0.7538	7544	7550	7555	7561	7567	7573	7578	7584	7590
0.83	0.7595	7601	7607	7612	7618	7623	7629	7635	7640	7646
0.84	0.7651	7657	7663	7668	7674	7679	7685	7690	7696	7701
0.85	0.7707	7712	7718	7723	7729	7734	7739	7745	7750	7756
0.86	0.7761	7766	7772	7777	7783	7788	7793	7799	7804	7809
0.87	0.7814	7820	7825	7830	7836	7841	7846	7851	7857	7862
0.88	0.7867	7872	7877	7882	7888	7893	7898	7903	7908	7913
0.89	0.7918	7924	7929	7934	7939	7944	7949	7954	7959	7964
0.90	0.7969	7974	7979	7984	7989	7994	7999	8004	8009	8014
0.91	0.8019	8024	8029	8034	8039	8043	8048	8053	8058	8063
0.92	0.8068	8073	8077	8082	8087	8092	8097	8101	8106	8111
0.93	0.8116	8120	8125	8130	8135	8139	8144	8149	8153	8158
0.94	0.8163	8167	8172	8177	8181	8186	8191	8195	8200	8204
0.95	0.8209	8214	8218	8223	8227	8232	8236	8241	8245	8250
0.96	0.8254	8259	8263	8268	8272	8277	8281	8286	8290	8294
0.97	0.8299	8303	8308	8312	8316	8321	8325	8329	8334	8338
0.98	0.8342	8347	8351	8355	8360	8364	8368	8372	8377	8381
0.99	0.8385	8389	8394	8398	8402	8406	8410	8415	8419	8423
1.00	0.8427	8431	8435	8439	8444	8448	8452	8456	8460	8464
1.01	0.8468	8472	8476	8480	8484	8488	8492	8496	8500	8504
1.02	0.8508	8512	8516	8520	8524	8528	8532	8536	8540	8544
1.03	0.8548	8552	8556	8560	8563	8567	8571	8575	8579	8583
1.04	0.8587	8590	8594	8598	8602	8606	8609	8613	8617	8621
1.05	0.8624	8628	8632	8636	8639	8643	8647	8650	8654	8658

$\mathrm{erf}^{-1}(1 - K_p)$	$(1 - K_p)$ 0	1	2	3	4	5	6	7	8	9
1.06	0.8661	8665	8669	8672	8676	8680	8683	8687	8691	8694
1.07	0.8698	8701	8705	8709	8712	8716	8719	8723	8726	8730
1.08	0.8733	8737	8740	8744	8747	8751	8754	8758	8761	8765
1.09	0.8768	8772	8775	8778	8782	8785	8789	8792	8795	8799
1.10	0.8802	8805	8809	8812	8816	8819	8822	8825	8829	8832
1.11	0.8835	8839	8842	8845	8848	8852	8855	8858	8861	8865
1.12	0.8868	8871	8874	8878	8881	8884	8887	8890	8893	8897
1.13	0.8900	8903	8906	8909	8912	8915	8919	8922	8925	8928
1.14	0.8931	8934	8937	8940	8943	8946	8949	8952	8955	8958
1.15	0.8961	8964	8967	8970	8973	8976	8979	8982	8985	8988
1.16	0.8991	8994	8997	9000	9003	9006	9009	9011	9014	9017
1.17	0.9020	9023	9026	9029	9031	9034	9037	9040	9043	9046
1.18	0.9048	9051	9054	9057	9060	9062	9065	9068	9071	9073
1.19	0.9076	9079	9082	9084	9087	9090	9092	9095	9098	9101
1.20	0.9103	9106	9109	9111	9114	9116	9119	9122	9124	9127
1.21	0.9130	9132	9135	9137	9140	9143	9145	9148	9150	9153
1.22	0.9155	9158	9160	9163	9166	9168	9171	9173	9176	9178
1.23	0.9181	9183	9186	9188	9190	9193	9195	9198	9200	9203
1.24	0.9205	9208	9210	9212	9215	9217	9220	9222	9224	9227
1.25	0.9229	9231	9234	9236	9238	9241	9243	9245	9248	9250
1.26	0.9252	9255	9257	9259	9262	9264	9266	9268	9271	9273
1.27	0.9275	9277	9280	9282	9284	9286	9289	9291	9293	9295
1.28	0.9297	9300	9302	9304	9306	9308	9310	9313	9315	9317
1.29	0.9319	9321	9323	9325	9328	9330	9332	9334	9337	9338
1.30	0.9340	9342	9344	9346	9348	9350	9353	9355	9357	9359
1.31	0.9361	9363	9365	9367	9369	9371	9373	9375	9377	9379
1.32	0.9381	9383	9385	9387	9389	9391	9392	9394	9396	9398
1.33	0.9400	9402	9404	9406	9408	9410	9412	9414	9415	9417
1.34	0.9419	9421	9423	9425	9427	9428	9430	9432	9434	9436
1.35	0.9438	9439	9441	9443	9445	9447	9449	9450	9452	9454
1.36	0.9456	9457	9459	9461	9463	9464	9466	9468	9470	9471
1.37	0.9473	9475	9477	9478	9480	9482	9483	9485	9487	9489
1.38	0.9490	9492	9494	9495	9497	9499	9500	9502	9504	9505
1.39	0.9507	9508	9510	9512	9513	9515	9517	9518	9520	9521
1.40	0.9523	9524	9526	9528	9529	9531	9532	9534	9535	9537
1.41	0.9539	9540	9542	9543	9545	9546	9548	9549	9551	9552
1.42	0.9554	9555	9557	9558	9560	9561	9563	9564	9566	9567
1.43	0.9569	9570	9572	9573	9574	9576	9577	9579	9580	9582
1.44	0.9583	9584	9586	9587	9589	9590	9591	9593	9594	9596
1.45	0.9597	9598	9600	9601	9602	9604	9605	9607	9608	9609

	$(1 - K_p)$									
$\text{erf}^{-1}(1 - K_p)$	0	1	2	3	4	5	6	7	8	9
1.46	0.9611	9612	9613	9615	9616	9617	9619	9620	9621	9622
1.47	0.9624	9625	9626	9628	9629	9630	9632	9633	9634	9635
1.48	0.9637	9638	9639	9640	9642	9643	9644	9645	9647	9648
1.49	0.9649	9650	9651	9653	9654	9655	9656	9658	9659	9660
1.50	0.9661	9662	9663	9665	9666	9667	9668	9669	9671	9672

[a] Abstracted by permission from W. Fish, *Methods Membrane Biol.*, **4**, 189–276 (1975).

[b] The values of $(1 - K_p)$ are given in the body of the table, while the corresponding values for $\text{erf}^{-1}(1 - K_p)$ are obtained from the row and column headings. For $K_p = 0.362$, $\text{erf}^{-1}(1 - K_p) = 0.645$.

APPENDIX III

Numbers for Testing Randomness of Groupings of Two Kinds of Objects at the 95% Confidence Level[a,b,c]

	m																		
n	2	3	4	5	6	7	8	9	10	11	12	13	14	15	16	17	18	19	20
2																			
3																			
4			3																
5		3	3	4															
6		3	4	4	4														
7		3	4	4	5	5													
8	3	3	4	4	5	5	6												
9	3	3	4	5	5	6	6	7											
10	3	4	4	5	6	6	7	7	7										
11	3	4	4	5	6	6	7	7	8	8									
12	3	4	5	5	6	7	7	8	8	9	9								
13	3	4	5	5	6	7	7	8	9	9	10	10							
14	3	4	5	6	6	7	8	8	9	9	10	10	11						
15	3	4	5	6	7	7	8	9	9	10	10	11	11	12					
16	3	4	5	6	7	7	8	9	9	10	11	11	12	12	12				
17	3	4	5	6	7	8	8	9	10	10	11	11	12	12	13	13			
18	3	4	5	6	7	8	9	9	10	11	11	12	12	13	13	14	14		
19	3	4	5	6	7	8	9	9	10	11	11	12	13	13	14	14	15	15	
20	3	4	5	6	7	8	9	10	10	11	12	12	13	13	14	14	15	15	16

[a] This table contains the minimum intergral numbers for groupings (runs) of m and n ($m \leqslant n$) objects of two kinds, for which the probability $P \geqslant 0.95$ for a random array. For example, the entry at at $m = 6$, $n = 10$ is 6, showing that any sequence of 6 objects of one kind and 10 objects of another kind containing 6 groups (runs) can be regarded as random at the 95% confidence level.
[b] Calculated probabilities as well as numbers for additional confidence levels can be found in F. S. Swed and C. Eisenhart, *Ann. Math. Stat.*, **14**, 16, (1943).
[c] Abstracted with permission from B. Mannervik, *Methods Enzymol.*, **87**, 373–390 (1982).

APPENDIX IV

Equations for Manually Calculating a Weighted Regression of a Linear Form of the Michaelis-Menten Equation.*

The initial estimates of K_m and V are given by

$$V = \frac{\sum v^4/[S]^2 \sum v^4 - (\sum v^4/[S])^2}{\sum v^4/[S]^2 \sum v^3 - (\sum v^4/[S])(\sum v^3/[S])}$$

$$K_m = \frac{\sum v^4 \sum v^3/[S] - \sum v^4/[S] \sum v^3}{\sum v^4/[S]^2 \sum v^3 - (\sum v^4/[S])(\sum v^3/[S])}$$

Once approximate values of K_m and V are obtained, more refined values can be calculated in a second iteration using the following equations.

$$V = \frac{(\sum v^2/(K_m + [S])^2(\sum [S]^2 v^2/(K_m + [S])^2)}{(\sum v^2/(K_m + [S])^2(\sum [S]^2 v/(K_m + [S])^2 -}$$

$$\frac{- (\sum [S]v^2/(K_m + [S])^2)^2}{(\sum [S]v^2/(K_m + [S])^2(\sum [S]v/(K_m + [S])^2}$$

$$K_m = \frac{(\sum [S]^2 v^2/(K_m + [S])^2)(\sum [S]v/(K_m + [S])^2) -}{(\sum v^2/(K_m + [S])^2)(\sum [S]^2 v/(K_m + [S])^2) -}$$

$$\frac{(\sum [S]v^2/(K_m + [S])^2)(\sum [S]^2 v/(K_m + [S])^2)}{(\sum [S]v^2/(K_m + [S])^2)(\sum [S]v/(K_m + [S])^2)}$$

* These equations were devised by A. Cornish–Bowden, *Fundamentals of Enzyme Kinetics*, Butterworth, London, 1979, pp. 178–179, according to the procedures recommended by G. N. Wilkinson, (*Biochem. J.*, **80**, 324–332 (1961).

This refinement process can be continued by additional iterations using the newly calculated values for K_m and V in the above equations. Iterations are continued until K_m does not change appreciably from one approximation to the next. Four iterations are usually sufficient. Computer programs for completing these calculations written for the Commodore 16K PET (Wilkin) and the IBM PC (Wilman3) computers are available for a slight charge from Instructional Media Center, Marketing Division, Michigan State University, East Lansing, Michigan, 48824.

Index